George Tragy-ovam

GW00640931

The Novice Licence
Student's Notebook

John Case, GW4HWR

38.50 £+ turne 12 pounds

Total Parts £38.50

Tuner + Knob additional £12.00.

Cheque payable to David Wood.
No Radis Club. 29th Jun.

Radio Society of Great Britain

Published by the Radio Society of Great Britain, Cranborne Road, Potters Bar, Herts EN6 3JE.

First published 1990
Second edition 1994, reprinted 1996

ISBN 1 872309 27 5

Cover design: Geoff Korten Design.
Illustrations: John Case, Ray Eckersley and Bob Ryan.
Production and typography: Ray Eckersley, Seven Stars Publishing.
Printed in Great Britain by Bell & Bain Ltd, Glasgow.

Contents

Preface

This notebook is intended to be used in conjunction with the Novice Licence training scheme. It cannot be used as an instruction manual to enable the reader to prepare for the examination without attending an approved course; in fact a licence can only be obtained by enclosing a course completion slip with the other paper work required. A course completion slip is obtained by candidates who have satisfactorily completed all aspects of the training course. THE COURSE IS AN ESSENTIAL PART OF THE QUALIFICATION.

Acknowledgement is made to members of the Training and Education Committee who have helped in the compilation of this book. We would also like to acknowledge the help given by several people in the preparation of the diagrams and photographic reductions etc.

John Case, GW4HWR

Part 1 Introduction

Hello! Welcome to the world of amateur radio. The fact that you are reading this suggests that already you know something about the Novice Licence. You may have read the introductory book, *Amateur Radio for Beginners*, or obtained information from a friend who is already a licensed amateur.

Whatever the reason, probably there are many questions you would like to ask, so here are the answers to some of them.

Who is allowed to have a licence for amateur radio?

Anyone who obtains the necessary qualifications can get a licence, but first you should know that there are a number of different types (categories) of licence:

- A full licence – Class A or Class B.
- A novice licence – Class A or Class B.

If you are 14 or older you can study for and obtain a full licence, either Class A or B.

The Novice Licence is available to students of any age, so if you are young it will be the one for you. If you left school many years ago we hope the new approach of the training for the Novice Licence will encourage you to have a go!

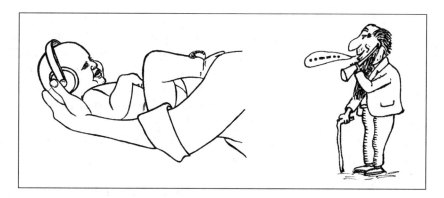

Of course you can try for the Novice Licence at any age and then move up into the other classes as soon as you are able to obtain the necessary qualifications – these are currently under review.

Tell me about the full licence.

As already mentioned, there are two types, known as 'Class A' and 'Class B'. Both require a pass in the Radio Amateurs' Examination (the RAE). To obtain an A licence you must also pass a Morse test at 12 words per minute.

Both A and B licences entitle radio amateurs to operate on the higher frequency bands known as 'VHF' and 'UHF'. 'VHF' stands for 'very high frequency' – these bands are similar to the VHF domestic radio band. 'UHF' stands for 'ultra high frequency' – bands like these are used by many television stations.

In addition they can use very much higher frequencies which are called 'microwaves'. These are similar to the frequencies used for satellite TV and also for microwave cooking!

The holder of an A licence can also use frequencies known as 'HF' which stands for 'high frequency'. You will know these as medium wave (where you can find the ordinary broadcast radio stations) and short wave, which you may have used to listen to radio stations all around the world.

What would I be able to do if I had a Novice Licence?

The Novice Licence also has two types A and B but the Morse test for the A licence is at only five words per minute. The licence allows the holder to use parts of selected bands, using transmitters which are limited in their power output. In spite of these limitations you would be able to make contact with stations in most of the countries in the world.

To be able to work like this you will need some patience and experience but many amateurs with full licences use only low-power (QRP) transmitters, sometimes with less power output than a bicycle lamp!

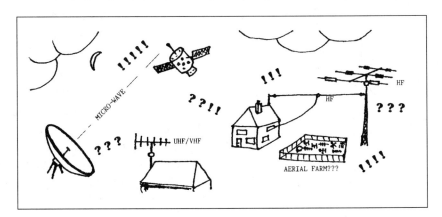

On the other hand you may wish to experiment with microwaves, or you may like to build a simple radio receiver which will allow you to take part in a 'game' called 'ARDF', which is short for 'amateur radio direction finding'. It is rather like a game of hide and seek. Each of the seekers has a radio receiver and uses it to find a transmitter which, together with its operator, is hidden in some convenient spot which cannot easily be seen.

Someone will need to build the simple transmitter; it could be you!

These are just a few of the ways in which the holder of a Novice Licence can enjoy the hobby of amateur radio.

How can I get a Novice Licence?

You must take part in a short course which will 'tell' you all you need to know in order to obtain the licence.

As this is a new licence, the training for it is also new and quite different from that of the Class A and B full licences. It is not possible to obtain the licence without attending an approved course.

Where are the courses held?

The courses are available throughout the UK and are held in many different types of club and group: radio clubs, Guide and Scout troops, schools and many others. Occasionally there may be a course which is held in the home of the course instructor. Get in touch with your Senior County Instructor; a list can be obtained from the Radio Society of Great Britain (RSGB) at Lambda House, Cranborne Road, Potters Bar, Herts EN6 3JE.

But what if I live miles away from the nearest group?

This can be a problem but don't give up hope! If there is a registered instructor living near you it may be possible to set up a course. Normally an instructor will look after a small group of four students so if there are any other people who might like to find out what amateur radio is all about, a group could be set up in your town or village.

Perhaps my parents won't let me go?

(This might be asked by a younger student). It is quite right that your parents are concerned, so suggest that one of them go along with you. If there is room in the group (remember, a maximum of four), perhaps they might like to join in the course – just for fun! Who knows, they may even get 'hooked' and end up with a licence – it has happened before. Even if there isn't room on the course, a parent could sit in and listen to the talks and watch the rest of you work.

Work!!! I thought this was supposed to be fun!

Of course it must be fun but like many hobbies there are things which must be learnt. The idea is to let you learn by doing as many things as possible rather than to sit and listen to the instructor talking. Of course some of the time she or he will have to talk so that you will know what to do.

Because the group is small the instructor will be able to give a lot of individual attention during the course. About half of the time will be spent on practical exercises. Most of the time you would be expected to work on your own with the instructor ready to help you.

You will be shown how to use radio equipment in the correct manner, tune in to stations in many parts of the world and get to know how to record your efforts. You will practise making calls using both speech and Morse code with the other members of the group acting as the 'other station'.

You will learn how to use tools to enable you to make up simple exercises such as test sets and amplifiers. You will also build your own radio and help to put up a simple aerial without hurting either yourself or anyone else.

Part of the time will be spent using meters to measure various things to be found in electronic circuits. Some of these you may have met in science lessons in school but if you have not (or it's so long ago that you've forgotten), don't worry, everything will be explained by your instructor. Although we keep saying 'instructor' you will find that she or he is much more a friend who will do everything possible to help you through the difficult bits.

This photograph shows three young people working on some of the course projects. Gavin has just completed his radio receiver which is described in Part 2. Abigail (centre) is examining a simple ARDF radio receiver and Karen is building the Test Set described in Worksheet 2. There is a 'deliberate mistake' in the picture; can you see what it is? If you can't, read Worksheets i and ii and look again.

Because the group is a small one you will soon know everyone and will speak to one another using first names. This is normal in amateur radio – 'Sir', 'Madam', 'Lord', 'Doctor' or 'Mr' etc are normally forgotten and very often not even known. Amateurs have been known to get into contact with a person who gives his name as Hussein and continue to use that name without realising that they are talking to a King!

It will not be long before you are ready to take the examination.

Examination???

Yes, there is an examination at the end of the course (City and Guilds NRAE 7730). The examination is taken at an approved CGLI centre. It is only a part of the qualification, the most important part being the work carried out during training. Almost every step in the work is assessed by the instructor and a licence can only be obtained after all the course work has been carried out correctly. The time spent in 'class' will be about 30 hours.

How do I apply for a licence?

This is the easy bit and is really what you will have been working for. When you have satisfactorily finished the course and your instructor has sent your Assessment Report form to the RSGB you will receive a Course Completion Slip. This will be required when you apply for your licence – it cannot be issued without it!

When you have passed the NRAE you will receive a CGLI Certificate of Unit Credit. This and the RSGB Course Completion Slip (which must pre-date the NRAE) will be sent, together with the completed licence application form, to the address shown on the form. If you are under 21 there is no fee but, if you are 21 or over, the appropriate fee should be included.

Your instructor will either provide a licence application form or advise you how to obtain one.

Then – perhaps the hardest thing of all – wait patiently for your licence to arrive!

Part 2

Make a medium-wave radio

Table 1. Components list

❏ Resistors:
R1, R5, R6	100k
R2	560R
R3	270R
R4	10k

❏ Capacitors:
| C1, C2, C3 | 100nF |
| Variable capacitor | 300–500pF |

❏ Semiconductors:
ZN414Z
BC548

❏ A crystal earpiece

❏ A 12-way, 2 amp terminal strip

❏ 22 metres of 0.4mm or 0.375mm diameter enamel-covered wire

❏ A few short pieces of coloured PVC insulated wire

❏ A miniature 3.5mm jack socket

❏ A 1.5 volt battery (AA cell) and box

❏ A toilet roll tube

❏ A few double-sided sticky pads

❏ Two drawing pins

❏ A soldering iron and a little cored solder

❏ A small terminal screwdriver

During the course you must make a simple working radio. This design, which we call the 'MF Rx', could be your project or you can make it in addition to another design. There is no fixed project for this part of the training but your instructor will advise you on your choice.

This design is very simple and can be built at home in an hour or so. With the exception of the tuning capacitor, the parts are all very easy to obtain and many of them will be found in someone's junk box. Most of the circuit is connected on an electrical terminal strip, the coil is wound on a cardboard toilet roll tube and the whole thing mounted on a block of wood.

To complete the radio you will need the parts shown in Table 1 and lots of patience! One of the attractions of this receiver is that it can be completed at home with just a screwdriver and a soldering iron.

A suitable tuning capacitor (300–500pF) may be a problem. They are rather expensive to buy but if you are able to visit a radio rally you will almost certainly find one for about one pound. You may find a tuning capacitor in an old LW/MW radio which has ceased to work. Your instructor will be able to help and may suggest that a two-gang capacitor connected together to make up a 500pF capacitor be used in the place of a single one.

1. Start by connecting the components on the terminal strip as shown in Fig 1, carefully checking the position and value of each one. The three capacitors are all the same and should not present a problem. The resistors are small cylinders with a wire coming from each end and the value will be coded by means of several coloured bands. Look at the first three bands – you should find:

Brown, black, yellow	= 100k	R1, R5 and R6
Green, blue, brown	= 560R	R2
Red, violet, brown	= 270R	R3
Brown, black, orange	= 10k	R4

It is most important to note the exact position of the IC (integrated circuit) marked 'ZN414Z' and the transistor marked 'BC548'. They look alike so check the numbers carefully.

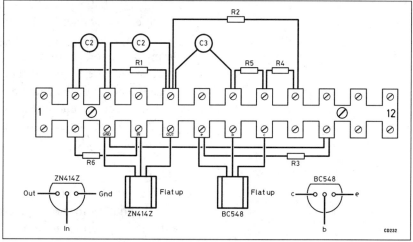

Fig 1. Terminal strip – position of components

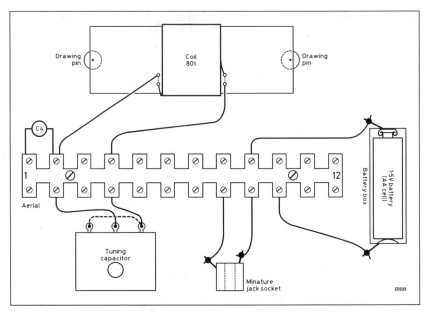

Fig 2. The layout of the parts on the wooden base

2. Now wind the coil. A standard toilet roll tube is about 42mm in diameter and 110mm long. Don't worry if yours is not exactly this size! Make two small holes in the tube about 40mm from one end. See Fig 2. Thread the enamel-covered wire through these holes, leaving about 100mm sticking out for the connections. Now wind on 80 turns of wire. Keep the turns close together but not on top of each other. This is one of the most important parts of the receiver so try to keep the winding tight. Be patient! When finished make two more holes and thread the end of the wire through them to hold the coil in position. If you have some clear varnish, give both the coil and the tube a coat and leave to dry.

3. The base can be any piece of soft wood about 150mm square and at least 10mm thick (an off-cut from a piece of floor board – supplied by a friendly builder perhaps). Look at Fig 2. Fix the terminal strip to the base with two small wood screws or, if you can't get any small enough, with double-sided sticky pads or even Blu-Tack™.

    Fix the coil near one edge of the base using two drawing pins and then connect the ends of the coil to the terminal strip as shown in Fig 2. Carefully scrape the enamel from the ends of the wire before putting them into the terminals. You will find it easier to remove the enamel if each end is held in the flame from a match for a moment or two – but please take great care – don't burn yourself!

    Fasten the battery box to the wooden base; small wood screws, double-sided sticky pads or Blu-Tack will do.

    Now you need six pieces of PVC insulated wire, two each for the tuning capacitor, the miniature jack socket and the battery box. (Use a piece of red wire for the battery box positive terminal, marked '+' and a black piece for the negative one, marked '−').

    To connect one end of each piece you will need a soldering iron. If you don't have one at home, this part of the job could be done when you are next in class.

Testing

4. Put the battery into the box (it must be the right way round – check Fig 2) and plug in the earpiece; you should be able to pick up some stations by altering the tuning capacitor. You may need to turn the whole radio to find the position which gives the strongest signal.

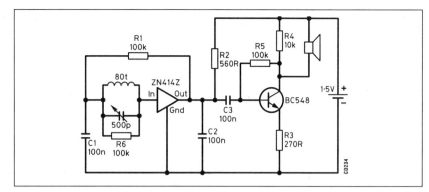

Fig 3. The circuit of the medium-wave radio

The receiver should work without any extra aerial but, if you live a long way from a medium-wave transmitter, try a long piece of wire hung up as high as possible and connected to terminal 1 on the connector strip. If you do this it will be necessary to connect another capacitor (another 100nF one) between terminals 1 and 2 on the connector strip.

A crystal earpiece will give a reasonable output. (No other type will work in this circuit).

When you have completed the audio amplifier project, the output of the radio can be fed into the amplifier and then everyone will be able to listen. Your instructor will show you how to connect the two units together.

The circuit of the radio is given in Fig 3 but you don't need to understand it or even to refer to it in order to complete the receiver; it is there for those who like to have a bit more information.

Your instructor will be able to help you to get the parts but, if there is any problem, almost everything can be bought from:

> Maplin Electronic Supplies Ltd,
> PO Box 3,
> Rayleigh,
> Essex.

They have branches in many big towns where you will be able to browse through the catalogue but, if you wish to order by post, you will need to buy one. They are normally on sale at W H Smith or of course their own shops.

Don't worry too much about the actual examination at this stage but concentrate on learning as much as possible in preparation for it. Listen carefully to the talks; refer to the worksheets to help in memorising what you have heard. Carry out the practical exercises and think about what you are doing and the results you have obtained. Ask your instructor if there are things which you don't understand.

Regularly look back at both the worksheets and the notes you have made. This applies to the practical exercises and the talks. If you have any doubts, make sure that they are cleared up during discussion times when they occur. There will be a number of these during the course.

After about 20 hours you can start thinking about the examination – it is one of those called 'multiple choice'.

Multiple choice???

Those of you still at school will probably know about this type of examination already. For those who do not, the following should help.

A question is asked by means of words or diagrams (or by a mixture of both). Four possible answers are suggested but only one of them is correct.

The candidate – that's you – is required to pick the answer which is most suitable for the question.

The following example should make the process a little clearer.

Question: Which one of the following Q-codes means "I am using low power"?

Answers:
a. QRL
b. QRO
c. QRP
d. QRT

All of these codes have some meaning but only one is correct – in this case it is c = QRP.

Don't be misled by suggestions that it is possible to pass the examination by guessing. It is true that you have one chance in four of guessing the correct answer to one question, but to guess the answers correctly to only 10 questions gives you one chance in more than one million. The examination requires that you should be able to give the correct answer to about 30 questions. If you think you can guess this number correctly you should try your luck at football pools!

How to pass

No book can really tell you how to pass the examination; this can only be done by means of your work during class and at home. Study and you will be able to find the answers. Even so there is considerable advantage in learning an examination technique. You will be able to practise this using one of the sample papers which will be available to you.

Go through the paper three times – call them the 'runs'.

First run

Start at question 1. Read the question very carefully. Next read the four 'answers' also carefully. If the answer appears to be fairly obvious, mark the answer sheet. (More about the method of doing so later.) If the answer does *not* occur to you almost immediately move on to question 2.

In other words, in your first run through the paper, only do the questions which appear easy. By the time you reach the end of the paper at question 45 you may well have 25 answers recorded correctly.

Each time you answer a question, and only if you answer it, put a tick against the number of the question on the question paper. This will stop you reading questions which you have already answered when you do the second run.

Second run

This time, look at the questions which you did not answer on the first run. Spend a little longer thinking about them, looking closely at the four options, eliminating them one at a time if you can't spot the correct answer. Even now, do not spend too much time on any one question – move on to the next and don't forget to mark the question paper whenever you answer any question.

Third run

Start at question 1 and check every answer you have given, then take a third look at the ones not yet answered.

Finally, as time is probably running out, make a guess at the remaining few questions not answered. No marks will be lost for wrong answers so choose one of the options – it might be the right one: one chance in four!

Recording your answers on the answer sheet

The actual method may vary from time to time and depends on the method used to carry out the marking. The method will be described on the paper and also the examiner will give you details before the start of the examination.

It is most important to be sure that the correct question is answered. Many examinations have been failed or almost failed because the candidate marked the answer sheet incorrectly.

If the answer to question 5 is recorded in the space for the answer to question 6, there is a good chance that the error will continue for a few questions before it is discovered. There will be a lot of corrections to be made, time will be wasted and there will be a considerable amount of panic.

Use a rule or a spare pencil to underline the correct position of the answer boxes. Fig 1 on the next page shows the general idea. If you are right-handed, place the answer sheet on the right-hand side of the question paper. If you are left-handed, the answer sheet should be on the left-hand side of the question paper and the following instructions modified to suit. Put the first finger of your left hand on the number of the question you are about to answer and keep it there until you have entered your choice in the same numbered box, using your right hand.

Don't forget to tick the number of the question on the question paper and then move on to the next question.

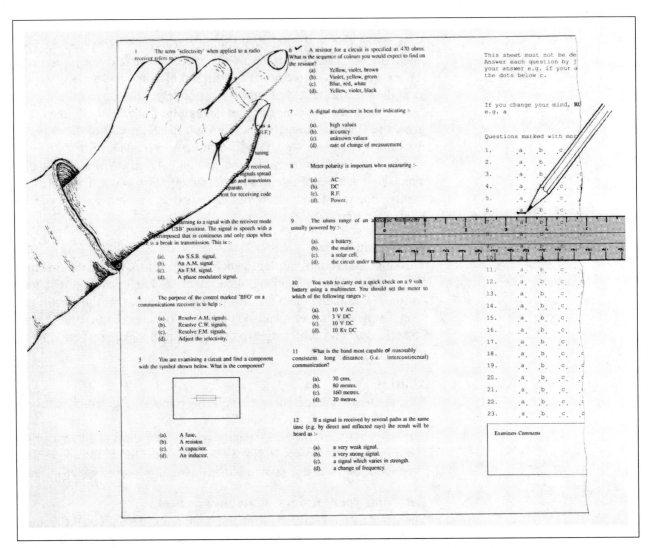

Fig 1. Underlining the correct position of the answer boxes

Part 4 **The Morse Test**

It will help with training for this part of the course if you read Worksheets 30 and 31, and carefully listen to the two tape recordings while you are in class.

The test consists of two parts – receiving and sending.

The receiving test

In this you will be expected to receive a short message in the form of a practical contact.

For example:

DL2XJK DE UA1VZP GA OM ES TNX FER CALL B̅T̅ UR RST IS 559 HR IN QTH LENINGRAD LENINGRAD ES NAME IS BORIS BORIS SO HW U CPY? A̅R̅ DL2XJK DE UA1VZP K̅N̅

You will find the meaning of the abbreviations in the above message in Worksheets 5, 6 and 31 but don't worry about them yet – there will be time for them later.

Although the speed of the message altogether will be only five words per minute, each separate Morse code character will be sent to you at a speed of 12 words per minute. There will be long gaps between characters. This may seem rather strange but actually makes things easier in two ways.

1. If characters are sent very slowly, they appear as a string of 'E's or 'T's – for example the letter 'A would sound like 'E' followed by 'T'.
2. Getting used to the sound of characters at the 12wpm speed will make it much easier for you to move up to a word speed of 12wpm.

The sending test

Again this will consist of a typical contact. Don't try to copy the character speed as used during receiving – concentrate on sending the correct dot/dash combination and with practice the correct spacing will be learnt.

Part 5

The worksheets

These have been prepared to help you do some of the work without too much help from your instructor.

Sheets have a letter following the number, either '(C)' or '(H)', and occasionally both. This indicates which sheets are intended to be used in class and which are used at home. Don't think of them as 'homework'. Because the course is rather short, you will not have much time to make as many notes of the things which your instructor tells you.

The sheets marked '(H)' will help to remind you of the things you have heard. Others contain details which you will need to remember and you should use these sheets as often as you can until all the items are remembered.

You will need to work to the sheets marked '(C)' when you are carrying out some of the practical exercises. For example, Worksheet 2 gives most of the information necessary to enable you to make up Test Set No 1.

Before you start building the project your instructor will make it up while you watch and she/he will point out some of the problems you might meet. Don't expect to be able to work as fast as your instructor – it is important to get it right even if it takes a long time.

Later the test set will be used for other things – for instance in the soldering exercise in Worksheet 8 you will use it to convince yourself that you have carried out the wiring correctly. Also, the test set is used with Worksheet 4 to enable you to measure some voltages and yet again when you get to Worksheet 12 while you discover for yourself some things about electricity. In these later exercises details of the test set will not always be given so that it will be necessary to refer back to Worksheet 2 to find out some particular detail.

While you are working on a particular exercise you may wish to make some notes. To help you to avoid getting them lost, the page opposite each sheet has been left blank. You can write your answers or other comments there. Try to keep your notes tidy and remember that things which appear obvious when you write them may not be so obvious when you look at them the next day or, even worse, the next week. Make your notes carefully so that you will understand them when you refer to them again.

As an example: '1.2V' doesn't tell you very much but if you had written 'voltage between points A and B equals 1.2V', and this appeared on the blank page to the left of Worksheet 4, it would be easy to see where the 1.2V had been found. Don't worry about the meanings of these notes at this stage.

Now look at some of the worksheets but please start with Worksheet ii – 'The Amateur's Code'. If we all follow this code the 'airways' will stay 'clean and tidy!'

INDEX OF WORKSHEETS

Notes

☞ *Syllabus sections 9.1–9.5*

When using some of the following worksheets you will be using tools, so here are a few simple rules which will help to avoid accidents which can happen if the rules are disobeyed.

When using the mains or batteries

Household mains voltage can easily produce a fatal shock, so never operate mains-driven equipment with the case open. If you need to look inside, switch off, disconnect the mains lead and then remove a panel and do what is to be done – fit a new crystal or something similar. Replace the panel and reconnect the supply. It is now safe to switch on.

Be sure that the fuse in the plug top is of the correct rating – a simple rule is that if there is no heating involved then a 3 amp fuse will normally be satisfactory. Your instructor will have already told you about residual current devices (RCDs).

Power from batteries is usually safe as the voltages are normally low. It is possible but very difficult to get a bad shock from a voltage of 30 volts or less. You need not fear the nine or 12 volts from 'dry' batteries of the type you will use to operate equipment. However, be very careful if a 'car type' battery is being used. It can only provide 12 volts but is capable of giving a lot of current which can cause wires to get red hot and cause fires or burn your hands.

When soldering

A hot soldering iron can cause a severe burn – it can also start a fire, so be sure to use a firm iron stand and return the soldering iron to the stand every time you finish using it. That will keep it away from yourself and the bench top. While you are using soldering equipment or tools, generally wear some protective clothing (overalls or other covering) to protect you and your clothes. If you have long hair, tie it back – especially when using a drill and, most important, when using power tools. If you are working on the kitchen table, you should protect the table top as well. A large piece of hardboard can be useful for this purpose. Get into the habit of wearing goggles – they may not look very good but your eyes must be protected.

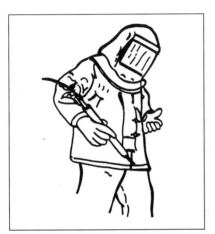

Wear protective clothing!

When using tools

Be very careful when using tools – hacksaws, files, screwdrivers, drills and wire cutters all seem relatively harmless but they can often inflict a nasty wound. Try to get into the habit of moving tools away from your body so that if the tool slips it will not hit you. Do the same thing when striking a match – if the head flies off it won't burn you or your clothes. Side cutters will sometimes cause a piece of wire to fly out when the end of a component is trimmed. Do not allow it to hit you or anyone nearby.

When working with steps and ladders

If you are helping to erect an aerial (or whenever someone is working above you) protect your head with a hard hat.

If using steps or ladders, be sure that they are secure and get someone to hold them if there is any doubt at all. Never use a ladder near any kind of overhead cables. Make sure that other people are warned of the danger of walking below someone working overhead.

Action in the event of electric shock

If someone near you receives an electric shock they may not be able to let go of the live conductor.

Notes

Do not touch them – you will almost certainly become a casualty yourself.

Look for the power point supplying the equipment and switch it off. Do not waste time – speed is of the utmost importance. When you are SURE that the supply is OFF you should check to see if the patient is breathing – call for help as loudly as you can. If you know how to carry out resuscitation, do so and keep trying until help arrives. If you do not know, any movement of the body will help – do something. Before the accident happens think about learning some first aid.

Notes

One

The Amateur is Considerate . . . He never knowingly uses the air in such a way as to lessen the pleasure of others.

Two

The Amateur is Loyal . . . He offers his loyalty, encouragement and support to his fellow radio amateurs, his local club and to the Radio Society of Great Britain, through which amateur radio is represented.

Three

The Amateur is Progressive . . . He keeps his station abreast of science. It is well-built and efficient. His operating practice is above reproach.

Four

The Amateur is Friendly . . . Slow and patient sending when requested, friendly advice and counsel to the beginner, kindly assistance, co-operation and consideration for the interests of others; these are marks of the amateur spirit.

Five

The Amateur is Balanced . . . Radio is his hobby. He never allows it to interfere with any of the duties he owes to his home, his job, his school, or his community.

Six

The Amateur is Patriotic . . . His knowledge and his station are always ready for the service of his country and his community.

Paul M Segal, ex-W3EEA, W9EEA

Notes

Silver 0.01

gold 0.1

☞ *Syllabus section 2(b)*

Table 1. The colour code

Black	0
Brown	1
Red	2
Orange	3
Yellow	4
Green	5
Blue	6
Violet	7
Grey	8
White	9

Gold and silver are also used for special purposes.

Different colours are used to mark the value on some components – this is often done with resistors, sometimes capacitors and once in a while, other components.

Ten different colours each represent a number which will always be the same for a particular colour. See Table 1.

There are a number of jingles to help remember the colours:

"Billy Brown Runs Over Your Garden But Violet Grey Wouldn't."

"Bye Bye Rosie Off You Go Blackpool Via Great Western."

The value of a resistor is usually given by three coloured bands but there may be one or two extra bands (these are not mentioned in the syllabus).

For now we will use the fourth and fifth bands (if they exist) to put the resistor in the correct position, as shown in Fig 1, so that the value can be read. Black or pink can never be the first band. Gold or silver can never be the second band, so if these bands are present turn the resistor so that they are on the right-hand end. If there are only three bands, there will be a bigger gap at one end of the body. In this case, turn the resistor so that this gap on the right.

The *first band* tells us the first figure of the value – for example, yellow would indicate '4'.

The *second band* gives the second figure of the value – for example, violet would indicate '7'.

So far we have '47'.

The *third band* tells us how many 0's (zeros or noughts) we must add – for example, orange represents 'three', so add '000' after the '47' to give 47000. That is the value of the resistor in ohms.

Watch out for a third band which is black, which means 'no zeros'. So a resistor having yellow, violet and black bands would be '47 with no zeros' or 47 ohms. Check with your instructor if you are not sure.

If the third band is gold, divide the number given by the first and second bands by 10. For example, yellow, violet and gold is 47 divided by 10, that is 4.7 ohms.

If the third band is silver, divide the number by 100. For example, 47 divided by 100 is 0.47 ohms.

You may discover that the third band is sometimes called the 'multiplier band'; don't worry about this but stay with the idea of 'how many zeros'.

Now look at each resistor in the box, write the colours in the correct order in your book and then write the value in ohms.

You may come across other ways of using the colour code and sometimes there may be six bands. In this instance the first three figures are given an actual value and the fourth band the number of zeros. This is not in the syllabus so there is no need to worry about it for now.

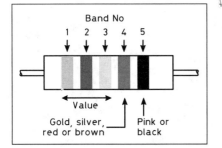

Fig 1. A typical coded resistor

Notes

First soldering exercise – Test Set No 1

☞ *Syllabus section 8*

The first objective in this exercise is to practise soldering and to make a basic test set. You will use this later for a number of different tests.

Check your materials:

- A piece of wood or chipboard a little larger than the outline in the diagram. This has been drawn actual size to make it easier to get the parts in the right places
- A bulb-holder
- A 2.5 volt, 0.2 amp (max) bulb
- Two battery boxes for AA size cells
- An 8.2 ohm, 0.25 watt resistor (grey, red, gold)
- A 2.2 ohm, 0.25 watt resistor (red, red, gold)
- Seven metal (brass) drawing pins
- About 300mm of tinned copper wire

You will also need a soldering iron, an iron stand, some resin-cored solder, a pair of wire cutters and a pair of 'long nosed' pliers.

Note – the name 'bulb' is used rather than 'lamp', which is correct, but go into a shop and ask for a 'torch lamp' and see what happens! Later you will find 'lamp' used in some of the exercises.

Construction

Look at Fig 2 and compare it with the parts.

The bulb-holder and the battery boxes may be in position already. If they are not, mount them in the positions shown, with very small screws or pieces of Blu-Tack™. Fix two solder tags under the terminal screws of the bulb-holder. Push the seven drawing pins into the positions shown –

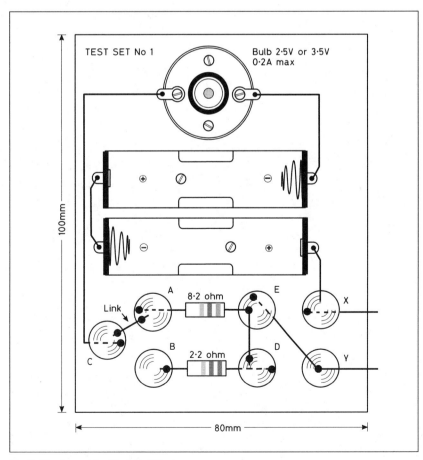

Fig 2. The layout for Test Set No 1

Notes

try to be as accurate as possible but do not push them down to the board at this stage!

Carefully solder pieces of wire and the two resistors to complete the circuit exactly as shown in the diagram but first read on.

The soldering iron must have a big 'bit' as the metal pins will cool a small bit and stop the solder flowing correctly. Make a start by 'tinning' the tops of the pins as you have been shown. If the solder doesn't flow smoothly, let the pin cool and then clean the top surface with fine wire wool. Now try again. Before you go any further, show the board to your instructor. You will soon discover that it is much more difficult to solder two wires to the same pin than it is to solder only one! The first one often falls off as the second is soldered. To stop this happening lead one wire under the pin and turn the end over so that it is on top. The pin can then be pushed down to the board. The end which you have turned over and the second wire can now be soldered in one go if you wish. *Note: the wire marked 'LINK' must not be trapped under pin A.*

Complete all the connections – the diagram shows the wires which should be trapped under pins. Show the completed board to your instructor and then carry out the following tests:

1. Put an AA cell (1.5 volts) into one of the battery boxes.
2. Carefully screw the bulb into the bulb-holder. *Do not try to keep turning when it stops.*
3. 'Short' (touch together) the two pieces of wire marked 'X' and 'Y'. What happens? When you have decided why – check with your instructor.
4. Follow his advice and try again. What happens?

If you are still unhappy about the result, check with your instructor again as you may have a bad soldered joint or have made a mistake.

Label the pins, using the same letters as those in the diagram. This will help you to carry out some of the future exercises which will refer to the letters.

Put your name on the underside of the board, using a sticky label so that it can be removed when you have finished all of the projects.

Notes

Make a log and design a QSL card

☞ *Syllabus section 6/6.6*

For some of the exercises you will need to fill up a log. Log books can be bought, but for now we are going to make our own. You can copy the table shown in the diagram but do not put in the entries which have been hand written. They are there to show you the type of entry that should be made. An amateur licence requires that a log should be kept but there are many reasons for keeping one even if it was not compulsory. It makes it possible to check on a station which you think you have worked before or look up the first name of the operator of the station you are about to work again. It gives a great feeling of friendship if you can say "Good evening Jean", or whatever the name happens to be, before the operator has time to remind you.

Look at the information regarding log keeping in the free RSGB booklet *A Closer Look at the Novice Licence*, page 4 – 'log book keeping'.

Now make out three sheets of log, keeping each page as neat as possible. Use a pen which does not smudge when you rub your hand over the lines, as you will be doing just that as you make your entries later.

When you make entries use a pen, not pencil, and there must not be any gaps – in other words every line must be filled in.

Requⁱred Optional

AMATEUR RADIO STATION LOG

DATE	TIME (UTC) START	TIME (UTC) FINISH	FREQ MHz	MODE	POWER	STATION	REPORT SENT	REPORT RCVD	QSL SENT	QSL RCVD	NOTES
30.4.91	15.10	15.30	14	CW	80W	AF2C	599	599	✓		AL NJ
"	15.20	15.45	"	"	"	K4SJ	559	579	✓		JOE FLA
4.5.91	16.00	16.20	"	SSB	"	W1ZAP	59	59	✓	✓	BILL MASS
"	16.30	16.35	"	"	"	CQ					
6.5.91	09.09	09.25	21	CW	"	UQ3ATE	559	579	✓		VLAD LATVIA
8.5.91	11.30	12.10	3.5	CW	"	G3JKS	599	599	✓	✓	FRED ST ALBANS
9.5.91	09.10	09.30	3.6	SSB	"	G0NDS	59	59	✓	✓	ROSIE HEMEL

Fig 3. An extract from a licensed amateur's log

You will already have been shown some QSL cards – now think about a design for your own card. It doesn't matter that you haven't a callsign as yet; just leave a space where it could be put in later. If you are a member of the RSGB (Radio Society of Great Britain), you will have been given a number which could be put into the card for the time being. You can design the card as a short wave listener (SWL) if you wish, but we think it would be best to start with a full card – it won't be long before you have your licence! If you belong to a club or group then why not put the name on the card?

Of minor interest, the card from VE7DEN in Fig 4 was the result of a contact with the author and Wally while on holiday on the Island of Elba in 1981. The author used the callsign GW4HWR/IA5. 'IA5' indicated that he was on one of the islands off the west coast of Italy. As you can see, Wally was on the western coast of Canada.

Notes

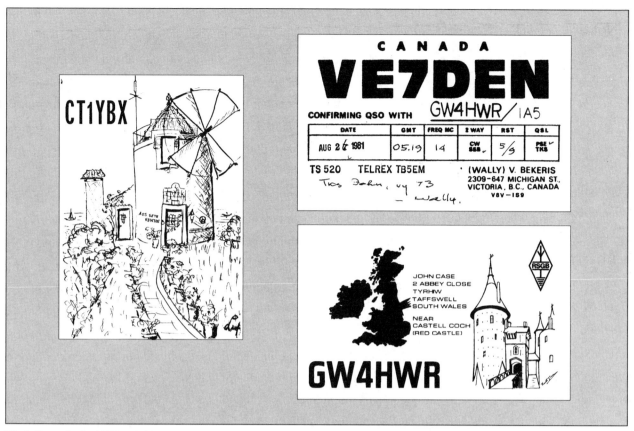

Fig 4. Some typical QSL cards

Notes

ANALOGUE

x,y TWISTED

$V_{BATTERY}$ = 1.6 V

V_{BULB} = 1.1 V

V_{CE} = 1.7 V

R

$R \div 50 \times Range$

x,y OPEN.

V_{XY} = E.00000000000 00000

DIGITAL

x,y Twisted

V Bulb = 1.05 V

VCE = 1.59 V

x,y open

V,x,y 2.82 V

Read and ENJOY

Putting a multimeter to work

☞ Syllabus sections 3(b)/3(c)

There are two types of meter in common use – *analogue* and *digital*. An analogue meter has a pointer which moves over a calibrated scale, while a digital meter displays the value of the voltage etc in figures (digits). Each type has advantages and disadvantages but let us examine both types and see for ourselves. Your instructor will already have shown you the meter/s – try to remember what was said as you watched and use the information. Fig 5 will help you to connect the meter correctly.

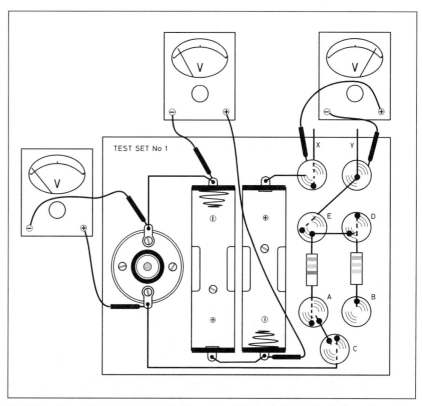

Fig 5. How the voltmeter is connected in some of the tests. Note: the indicated position of the pointers may not be correct

Analogue meter

This will have one or two switches to enable the meter to be set as a voltmeter, an ohmmeter or as an ammeter and also to fix the correct range. Try setting the meter to different ranges and modes (voltage, current or resistance) and examine the scale very carefully.

Set up Test Set No 1 with two AA cells inserted into the battery boxes. Short X and Y and check that the bulb lights. If it doesn't, look carefully at the circuit – you should be able to see what is wrong. Now set the multimeter to measure DC (direct current) voltage and choose the best range. You should be able to do this without any help as you probably know the total voltage of the batteries.

Measure the voltages *between* the two terminals of each battery. If the pointer tries to move in the wrong direction you have connected the meter incorrectly – think, then try again.

Now check *between* the two terminals of the bulb-holder and also between C and E. In each case try to work out the voltage indicated by the pointer on the scale. You will soon get used to it – just keep trying. Write down what you consider is the voltage in each case and check with your instructor.

Remove the shorting link between X and Y and then test the voltage

Notes

between X and Y. Does the bulb light? Is there any voltage? If you don't understand these last two effects, ask your instructor.

Digital meter

These are often very clever devices which will set themselves to the most suitable range ('auto ranging'), so all you have to do is to set the mode, that is, tell the meter to measure voltage. The figures will sometimes contain a decimal point: this is read in the usual way. After the figures you may find some letters such as 'mV' or 'V'. The first stands for 'millivolts'. One volt is equal to 1000 millivolts; in other words a millivolt is a thousandth part of a volt. Later you may see 'kV', standing for 'kilovolts'. A kilovolt is 1000 volts but you won't be using voltages like this for a long time! You will, however, meet 'kΩ' which stands for 'kilohm' or 1000 ohms and 'MΩ' ('megohm') which is 1,000,000 ohms.

Now measure the same voltages as in the first part of the exercise. Easy, isn't it, so what's the snag? When first developed, digital meters were much more expensive but are now reasonably cheap. You will probably have noticed that you have to wait for the reading to stabilise. The first reading you get may not be correct and many people find this delay rather annoying – it will make it impossible to carry certain types of test. In practice it would be very nice to have one of each type of meter!

Now measure all the voltages between the points indicated above, this time using the DVM (digital voltmeter). Compare them with the values you got using the analogue meter.

Add the voltages of the two batteries together. Add the voltage across the bulb to the voltage between C and E and look at the two totals. What do you notice?

Notes

Codes and abbreviations

☞ *Syllabus sections 6(c)/6(d)*

Somebody calls!

Strong

Weak

I'll call you back

Most of the following codes etc are for use when you are working with Morse code but you will sometimes hear them when telephony (speech) is in use by a station. Try to avoid falling into this habit – there is no advantage in saying "I am going to go QRT" instead of "I am going to close down". The second is the correct way when using 'phone' (speech). Occasionally it will be quicker to use a Q-code and sometimes it will be quite acceptable. We will point out some of these as they occur in the following paragraphs.

The Q-code

This is an international code used widely by many operators in civil and military ships and aircraft. It has been adopted by radio amateurs and sometimes the meaning has been slightly altered. Here are some in common use:

QRA – The name of my station is . . .
QRG – Your exact frequency is . . .
QRK – The intelligibility of your signal is . . .
QRL – I am busy (This frequency is in use).
QRM – Interference (from other stations).
QRN – Interference (from thunder storms or nearby electrical apparatus).
QRO – High power.
QRP – Low power.
QRQ – Send faster.
QRS – Send slower.
QRT – Close down.
QRX – Stand by, I will call you again. (This is useful while using phone when something happens which must be attended to quickly).
QRZ? – Who is calling me?
QSA – The strength of your signal is . . .
QSB – Fading. The signals change in strength due to the conditions.
QSL – Confirmation of a contact. QSL cards are sometimes exchanged.
QSO – A radio contact.
QSY – Change frequency.
QTH – The place from which you are operating (location).

When a statement becomes a question the Q-code must be followed with a '?' (di-di-dah-dah-di-dit). An example is that 'QRQ' means 'Send faster' but 'QRQ?' means 'Shall I send faster?'

Abbreviations

There are over 100 abbreviations used by radio amateurs when making calls in code but for the beginner it is better to make use of just a few of them. You should try to remember the following:

CQ   – This is used in both telephony and Morse code contacts and means 'Calling all stations'. Your licence will allow you to make this general call only when you are trying to set up a contact.

CW   – Short for 'continuous wave' – commonly used to describe a Morse code transmission.

DE   – Means 'from' and is put between the callsign of the calling station and the callsign of the called station. Example: G3PFR DE GW4HWR.

K   – 'Carry on', an invitation to anyone to respond to your call.

R   – 'Message received'.

$\overline{AR}$ – (A bar over the letters indicates that they are sent as one character). Di-dah-di-dah-dit. Means 'end of transmission'.

$\overline{SK}$ – Di-di-di-dah-di-dah – means 'end of work' and indicates that the QSO
or $\overline{VA}$ is finished and other stations may now call.

Notes

$\overline{BT}$ – Dah-di-di-di-dah is used to mark separate parts of the same transmission. (=)

MSG   – Message.

BK   – Signal used to interrupt a transmission in progress.

WX   – Weather.

PSE   – Please.

RX   – Receiver.

TX   – Transmitter.

SIG   – Signal/s.

UR   – Your.

YL   – Young lady. This is used during both phone and Morse code transmissions.

YM   – Young man. Young people may like to use this in the place of the more traditional OM (Old man).

73   – Best wishes. *Note*: not 73's or VY 73's. Although you will hear this during 'phone' contacts, it is much better to say "I wish you all the best" or something similar. Keep 73 for code transmissions.

88   – Love and kisses. Again, not 88's. Perhaps it might be safer to use this for both 'phone' and Morse code messages!

Worksheet 31 includes abbreviations which need to be learnt before the 5wpm Morse test is taken (in order to obtain a Novice Licence A).

Notes

More codes and abbreviations

☞ *Syllabus sections 6(d)/6(e)*

In preparation for short-wave listening, a knowledge of the following is necessary to understand what is being 'said' on air.

The RST code

This is the way we tell one another about the signals we receive.

R for readability

R1 'Unreadable' – the received signal is so bad that it is impossible to 'read' a single word.

R2 Barely 'readable'. Occasional words can be picked out.

R3 'Readable' but with considerable difficulty.

R4 'Readable' with almost no difficulty.

R5 Perfectly 'readable' – every word can be picked out.

S for signal strength

S1 Faint signals – almost inaudible.

S2 Very weak signals.

S3 Weak signals.

S4 Fair signals.

S5 Fairly good signals.

S6 Good signals.

S7 Good signals – moderately strong.

S8 Strong signals.

S9 Very strong signals.

When using telephony, only the R and S values are given. For example the station you are working with may tell you "Your signals are 4 and 7". This means that almost every word has been heard correctly and the signal is strong enough to blot out most of the background noise.

The value of the signal strength can sometimes be read from a meter, known as the 'S-meter', but these meters are not usually very accurate and it is best to judge the strength by listening to the overall sound.

When using Morse code an additional piece of information is given – the quality of the note of the signal, known as the 'tone'.

T for tone

T1 Extremely rough hissing sound.

T2 Very rough note, not at all musical.

T3 Rough low pitched note – slightly musical.

T4 Rather rough but moderately musical.

T5 Musically modulated note.

T6 Modulated note with a slight trace of whistle.

T7 A musical note with some ripple.

T8 A very good note with just a trace of ripple.

T9 A pure musical note.

When you are working a station using Morse code you may receive:

= UR (your) RST 579 =

This indicates that every word has been received, the signal strength is very good and the quality of the audio note is excellent.

Note that '=' is the way the break signal $\overline{BT}$ is written.

The international phonetic alphabet

When working a station using telephony it is sometimes necessary to spell some words but it is of little use just spelling them as many letters sound very much the same, especially if the signals are not too strong or are noisy. So we make use of the alphabet listed on the next page.

Notes

Letter	Word	Pronounced
A	Alpha	AL FAH
B	Bravo	BRAH VOH
C	Charlie	CHAR LEE
D	Delta	DELL TAH
E	Echo	ECK OH
F	Foxtrot	FOKS TROT
G	Golf	GOLF
H	Hotel	HOH TELL
I	India	IN DEE AH
J	Juliet	JEW LEE ETT
K	Kilo	KEY LOH
L	Lima	LEE MAH
M	Mike	MIKE
N	November	NO VEM BER
O	Oscar	OSS CAH
P	Papa	PAH PAH
Q	Quebec	KEH BECK
R	Romeo	ROW ME OH
S	Sierra	SEE AIR RA
T	Tango	TANG GO
U	Uniform	YOU NEE FORM
V	Victor	VIK TAH
W	Whiskey	WISS KEY
X	X-Ray	ECKS RAY
Y	Yankee	YANG KEY
Z	Zulu	ZOO LOO

Use the phonetic alphabet when conditions are poor, when speaking to someone who does not normally use English or when a strange name is used. For example "Llanelli" would be difficult (if it was pronounced correctly) and the other station was being operated by, say, a German-speaking person but "LIMA, LIMA, ALPHA, NOVEMBER, ECHO, LIMA, LIMA, INDIA" is more likely to be received correctly.

If you do use the phonetic alphabet it is most important to use the internationally agreed words – they should be understood by amateurs everywhere.

Finally

When using telephony, try to avoid the use of Q-codes – keep them for Morse code contacts. It is much better to say "Send my best wishes to John" rather than "QSP my 73 to John". Generally speaking, good English is easier to follow than a mixture of 'roger' and 'affirmative' and many others.

The author has been asked several times by the operator of a foreign radio station "What is the meaning of the English word 'offenclear'?" Almost certainly the words heard were "off and clear". This would not be understood easily by someone with a little knowledge of English. "This station is closing down" would have been much better.

☞ *Syllabus section 6(a)*

This practice must not take place over the air.

For this exercise two students will work together but the plan is arranged so that in the group of four, each student will work with the other three as the scheme progresses. Ideally a pair of telephones in separate rooms would give the best feeling of an actual contact on air. Two old telephone handsets connected together by means of some twin cable with a battery in series is all that is needed. If this is impossible then the contact must be carried out by the two 'operators' just talking to one another. If the materials are available but the system is not set up, the first two operators could spend a little of the time helping the instructor to do so.

By this time you will have prepared some log sheets, and these can be used to record your contacts.

First decide who is to listen and who is to make the initial call and then make up some callsigns for each of you. You should use the correct Novice-type callsign which will look something like this: '2E0XXX'. The '2' indicates a Novice callsign, the 'E' that the operator is in England, the '0' that it is a Class A licence and the 'XXX' is the personal part of the callsign.

To help you to make up some callsigns, here is some additional information. A Novice call always starts with '2'. The second symbol is a letter which indicates the country in which the operator is working.

E	England
W	Wales
M	Scotland
I	Northern Ireland
U	Guernsey
J	Jersey
D	Isle of Man

The third symbol is either '0' or '1' indicating the class of the licence (A or B).

Make up your callsigns and decide who is to listen and who is to make the initial call.

When you are both ready the calling station can make a start by listening for about 30 seconds and then saying "Is this frequency in use please?" Repeat: "Is this frequency in use?" This is necessary because when two stations are in communication you may not be able to hear one of them. You will learn the reason for this later. If there is a contact in progress the station which is listening will answer with "Yes, the frequency is in use." In practice you would then look for another apparently free frequency.

If there is no reply to your question the calling station can make the initial call:

"CQ CQ CQ. This is 2E0XYZ calling. CQ CQ 2 ECHO zero XRAY YANKEE ZULU, 2 ECHO zero XRAY YANKEE ZULU calling CQ and standing by."

Repeat if necessary. The calling station will record this in the log, together with the other required information. An example of the log entry has been given in the worksheet about making log sheets.

The receiving station should enter the call in the log and then answer by saying:

"2 ECHO zero XRAY YANKEE ZULU, here is 2 WHISKEY zero LIMA ECHO FOXTROT, 2 WHISKEY zero LIMA ECHO FOXTROT – 2W0LEF standing by."

The call will be entered in the logs of both stations for the purpose of this exercise.

Note that both the calling station and the receiving station use the

Notes

international phonetic alphabet for the initial calls. This is done so that both stations can be sure that the other has received the callsign correctly. Once this has been done there is no need to use phonetics when giving the callsigns. Although the licence requires the callsign to be given only at the commencement and end of each contact, or at intervals of 15 minutes if the contact is a long one, it helps the flow if your callsign is given at the end of each 'over' as it indicates that the call has been handed over.

2E0XYZ should continue with:

"2W0LEF, this is 2E0XYZ, thank you for the call. My name is . . . and my location is . . . Your signals are five and nine." (Or whatever you think is justified; if someone in the room is making a lot of noise it may be difficult to hear and then you could say "your signals are three and seven".) "How are you receiving me? 2W0LEF from 2E0XYZ."

Both stations make the necessary entry in their logs.

2W0LEF should now make a similar call, giving the signal report and location of the station. Each call is entered in the logs. If there is nothing further to say, 2W0LEF could finish by saying:

"Thank you for the call, I look forward to hearing you again. My best wishes to you. 2E0XYZ, this is 2W0LEF signing with you."

2E0XYZ will complete the contact with a similar 'thank you and best wishes' and then:

"2W0LEF, this is 2E0XYZ closing down." (or changing frequency or standing by).

Every call is to be entered in the log of both stations. Show your logs to your instructor to make sure that you have correctly recorded the contact.

If time permits, set up a new call in the same way. Do not try to talk while the other station is speaking. Normally this would be impossible in practice and it may be that your instructor has modified the handsets. In this case you will need to operate some kind of switch (referred to as a 'press to talk' or 'PTT' for short) in order to make your 'transmissions'.

Keep your 'overs' short and spend as much time as possible using the 'change over' techniques. When you pass over the transmission, always put the other person's callsign first.

Your transmissions may contain information about yourself, the type of station you hope to have, the weather and details of the town or village in which you live. You must not talk about politics or try to sell any goods etc. Your instructor will give you more information.

Notes

Worksheet 8 (C)   **Soldering exercise No 2**

☞ *Syllabus section 8*

The object of this exercise is to practise soldering on a printed circuit type board and to follow simple instructions in words and pictures.

The materials and tools required are:

- A small piece of Veroboard with at least six copper tracks
- A few pieces of tinned copper wire, soldering iron with a thin bit, soldering iron stand and solder
- A pair of wire cutters and long-nosed pliers

Examine the Veroboard to see how the holes and copper strips are arranged. Now look at Fig 6.

Any hole can be named by using the hole letter and the track number. For example, the hole with the ring round it is C4.

Find A1 and A5. Take a piece of wire about 20mm long and bend it into this shape ⌐ so that the two points will fit nicely into the two holes with the wire flat against the plain side (component side) and the points sticking out at least 1mm on the track side. Carefully solder these points to the track. Cut off surplus wire and then show it to your instructor. If OK, link the following holes in the same manner: B2 to B4, then C1 to C4 and finally D2 to D6. Solder all links at each end as they are fitted and cut off surplus wire.

Use Test Set No 1 to check between E5 and E6. Touch the wire at X on Test Set No 1 to E5 and wire Y to E6. What happens? Test between E3 and E6. What happens? Report both results to your instructor.

The checks you have made are called 'continuity tests'.

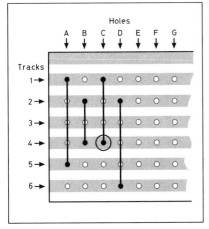

Fig 6. A simple Veroboard exercise

Notes

☞ *Syllabus sections 4(a)/4(e)/4(f)*

Please don't be put off by the strange sounding name – 'propagation' is just a posh way of saying 'the way in which the radio signal gets from one place to another'.

Radio signals or waves are almost exactly the same as light waves. The only real difference is the frequency at which they occur. If you shine a torch up into the air on a misty night you will be able to 'see' the beam as the light waves light up the particles in the air, but if the beam shines on to a wall there will be nothing beyond the wall. If a mirror is placed in the right position it will be possible to make the light beam go over the wall by shining it on to the mirror. The same sort of thing occurs with radio waves.

A transmitter sends out a wave which may travel parallel to the ground, but because the surface of the earth is a curve, the wave will tend to move away from the surface as it gets further from the transmitter. At lower frequencies, such as those used by the long- and medium-wave broadcast stations, the wave will bend a little to follow the curvature of the earth. This wave is called the 'ground wave' and during daylight hours is the signal which we use to listen to our favourite stations on the medium- or long-wave bands. Because the wave is in 'contact' with the earth, much of the energy is absorbed and the wave does not carry too far. As the frequency of the wave gets higher, so the absorption increases and the distance travelled gets less.

So how do radio waves travel long distances? We need a mirror and we have one. It used to be called the 'Kennelly-Heaviside layer' but has since been shown to consist of a number of different layers which are together called the 'ionosphere'. These layers have the ability to bend some waves so much that they return to earth a long way from the point where they started. Fig 7 will make this easier to understand.

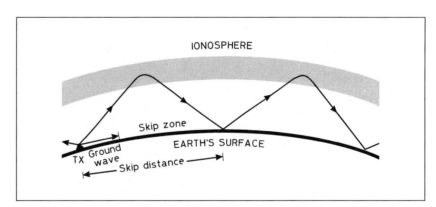

Fig 7. Reflected waves can travel around the earth's curved surface

The distance from the transmitter to the point where the signal first returns to earth is called the 'skip distance' and the space between the end of the ground wave to the point where the waves comes back to the earth's surface is called the 'skip zone'. A receiving station in the skip zone will not be able to receive the transmission. Very often the signal will be reflected by the earth and carry on further around the earth on a second (or third) hop. In this way we will be able to receive signals from the opposite side of the world.

Generally it is the long, medium and short waves which will be 're-flected' in this way. Very high frequency (VHF), ultra high frequency (UHF) and micro waves are not normally 'bent' enough and do not come

Notes

Novice frequency bands and distances that can can be worked		
Band	**Frequency (MHz)**	**Notes**
160m	1.950–2.000	'Top band'. Semi-local signals during the day, more distant signals after dark. In winter during early hours long-distance signals can be heard.
80m	3.560–3.585	Local and western European stations can be heard with long-distance stations being present during the evening in winter.
30m	10.13–10.14	Mainly European stations heard during daylight hours but long-diatnce stations heard at other times.
15m	21.100–21.149	Best in spring and late autumn at the peak of the sunspot cycle. Tends to fade out after dark. Intercontinental signals can be heard during the day.
10m	28.060–28.190 28.225–28.300 28.300–28.500	At height of sunspot cycle, signals can be heard from all over the world from morning to night. Often good over north/south paths – sporadic-E propagation occurs. During low sunspot activity the band appears very quiet but occasional DX can be heard.
6m	50.0–51.0 51.0–52.0	Local regional signals are heard. At peak of the sunspot cycle the east coast of the USA can be heard.
70cm	432.0–440.0	UHF band where local contacts can be made. Distances of many hundreds of miles can be worked under certain circumstances.
23cm	1240–1325	Line-of-sight communications. A band for the experimenter.
3cm	10,000–10,500	Line-of-sight communications. Another band for the experimenter.

back to earth. With the last three frequencies mentioned, we rely on the ground wave and can only receive stations from a relatively short distance as with VHF radio and UHF television. To get the greatest distance we try to put both the transmitter aerial and the receiver aerial as high as possible.

Just as the wall will stop the light from the torch, solid objects such as buildings and trees will reduce the strength of the higher-frequency signals and we say that they travel 'line of sight' – in other words, as far as you can see (plus about a third under 'normal' conditions).

Unfortunately the reflecting layers are affected by a number of different things – the time of day, time of year and the sunspot activity. This changes considerably and has an approximate 'cycle' of about 11 years. When the activity is high, radio communication in the HF bands improves considerably and in the trough of the cycle, tends to be rather poor. Ask your instructor to tell you the present state of the sunspot cycle. You will remember that 'bands which are open' was a term used when listening. It is the changes in the reflecting layers which causes these effects.

On VHF and UHF the weather often produces surprising effects and long-distance communication may suddenly be possible when the weather conditions are 'right'. This often occurs during the state referred to in the weather reports as 'high pressure'.

Notes

A look at aerials

☞ *Syllabus section 4*

Radio amateurs make use of a wide variety of aerials, each one having certain properties which make it suitable for a particular application. In this exercise we will look at just two types – the dipole, widely used for both receiving and transmitting, and the ferrite rod, used almost exclusively for receiving.

Examine the dipole. If there are several rods mounted on a central, larger rod (known as the 'boom'), look for the one with the cable connected to the centre. That is the dipole (or driven element), the other rods are 'directors' and 'reflector'. More about these later. The complete aerial is called a 'Yagi'! Look at the way in which the cable is connected to the dipole. It will probably be coaxial cable. Note that there are two connections: one is the inner conductor with PVC insulation and the other is the outer braid which is covered by an outer sleeve. Pay special attention to the way in which these two conductors are connected to the dipole as you will probably have to connect an aerial like this in the future.

Measure the length of the dipole in centimetres, add five percent of the length (multiply by 1.05) and then multiply by two. The result is the approximate wavelength of the signal on which the aerial is designed to operate. If there are rods other than the dipole, measure them. The one which is longer than the dipole is the 'reflector' and the shorter one/s are 'directors'. If there are several directors note the length of each one – is there any pattern?

Now connect the dipole to a receiver. If possible, use one which has a signal strength meter (S-meter). Set the receiver to the band indicated by your calculation above. Listen for a signal which is fairly constant, then turn the dipole in various directions and note how the signal varies. The receiver may work better when the dipole is vertical, if so the signal is said to be 'vertically polarised' but if the receiver works best with the dipole horizontal, it is 'horizontally polarised'. If the first is true, does the receiver work at all with the dipole horizontal? If the answer is "yes", try turning the dipole in a circle and describe the result. If the signal is vertically polarised and the dipole has one or more directors, try turning the aerial in a circle while it is in the vertical position. Describe the result.

Next look at the ferrite rod aerial. This will probably be part of a portable transistor radio. Describe what you can see on the rod. Can you say what these things are? If in doubt, ask your instructor.

Switch on the radio using either MW or LW, tune in a station and note the effect of turning the radio. Can you make a rule which will enable you to put the radio in the best position to receive a station? Again check your ideas with your instructor.

If you would like to learn more about aerials you will find the RSGB book *Practical Antennas for Novices* by John Heys, G3BDQ, very helpful.

Notes

Audio frequency amplifier

☞ *Syllabus section 8*

This is a useful amplifier which can be used with many other projects, such as a crystal radio or the MF receiver, and will convert it into quite a 'noisy' device. It is most important that you do not rush to get the amplifier finished – concentrate on doing your best work.

Make a start by examining the components – you should have:

- One PCB. The plain side is called the 'component side' and the other the 'track side'. Fig 8 shows the track side full-size.

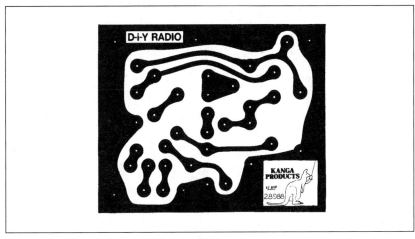

Fig 8. The foil pattern of the PCB – track side

- Three resistors. Look for a gold or silver band – turn the resistor so that this band is on the right-hand side. Now look at the three coloured bands at the left-hand end. Yellow, violet, red stands for 4700 ohms or 4k7. Brown, grey, brown stands for 180 ohms or 180R. Brown, red, green stands for 1,200,000 ohms or 1M2.
- Four capacitors. The two small 'beads' are tantalum capacitors and will be marked '4.7μF' or '4μ7' with a '+' sign above one lead. One other is a large tube with a wire coming from each end (axial). This will be marked '220μF' with a '+' or a '−' at one end. The remaining capacitor is a tube with two wires coming from the same end (radial) and this will be marked '47μF'. These are said to be 'polarised' and must be connected the correct way round, so it is important to find those '+' or '−' signs.
- Two diodes. These are small glass things with a black band at one end and may be marked '1N4148'.
- Three transistors. Two of these will be marked 'BC548' or 'BC182' and the other 'BC558' or 'BC212'.
- One potentiometer (volume control).
- One loudspeaker – be careful with this and don't let anything press against the cone.
- One PP3 battery clip with two leads, one red and the other black.

Important – In some kits the capacitors C1 and C2 may be replaced by ceramic types with a lower value and somewhat smaller. These will work just as well and will fit under the volume control (potentiometer) with less trouble. They are *not* polarised and can be fitted either way round. The volume control may be marked '47k log' instead of '25k log'. Again this will be perfectly satisfactory.

Lay the PCB on the track side so that the 'D-i-Y RADIO' sign is at the top but hidden from view. Now compare the holes with the dots in Fig 9.

Notes

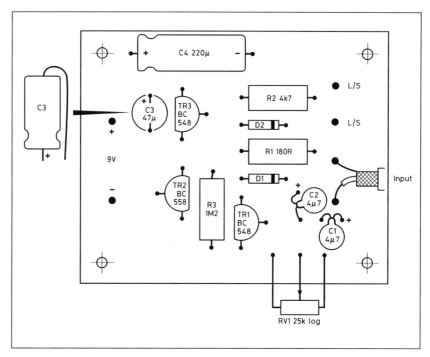

Fig 9. Position of the components on the printed circuit board (PCB)

Bend the wires of the three resistors at right-angles to the body (do not use pliers) so that they fit the holes in the board. When you are sure that they are in the right holes and that the value is correct, push the resistors down so that they lie flat on the board. When you are satisfied that everything is correct, bend the wires apart on the track side of the PCB (just enough to stop the resistors falling out when the board is turned over) and solder each of the wires to the PCB – cut off the surplus wire. If you are good at soldering you may cut the wires before soldering them.

Now fit the four capacitors. These must be connected the right way round – note the '+' and '−' signs on the component and on the drawing in Fig 9. Again the components should be close to the board, not standing up on stilts!

Note that C3 is mounted with the leads down. They are shown on top in the diagram to ensure that the positive one is put into the correct hole.

If an axial-type capacitor is supplied for C3 it must be fitted 'on end' – see the inset in Fig 9.

The two diodes can be fitted next but make certain that they are the correct way – note the band on the components and in Fig 9. Try to solder these components quickly as they can be damaged by excessive heat.

The three transistors should be mounted so that the body is about 5mm above the surface of the PCB. It is most important that the wires of each transistor go into the right holes and that the correct type is used in each position. If the flat part on the body is in the same position as shown in Fig 9, there should be no problem. Fit the control so that the spindle sticks out from the front edge of the board.

Connect a piece of red insulated wire to the pad marked '+9V' and a black piece to the pad marked '−9V', then solder these to the switch on the back of the control. Connect the two leads from the battery clip to the other two tags on the switch as shown in Fig 10.

Finally connect the loudspeaker using two pieces of insulated wire about 100mm long and twisted together.

Later you may wish to put the amplifier into a box. There is no problem; almost any box that is big enough will do. All that is needed is one hole big enough to accept the bush on the control. Pass the control through

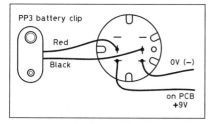

Fig 10. Connections to the switch on the back of RV1

Notes

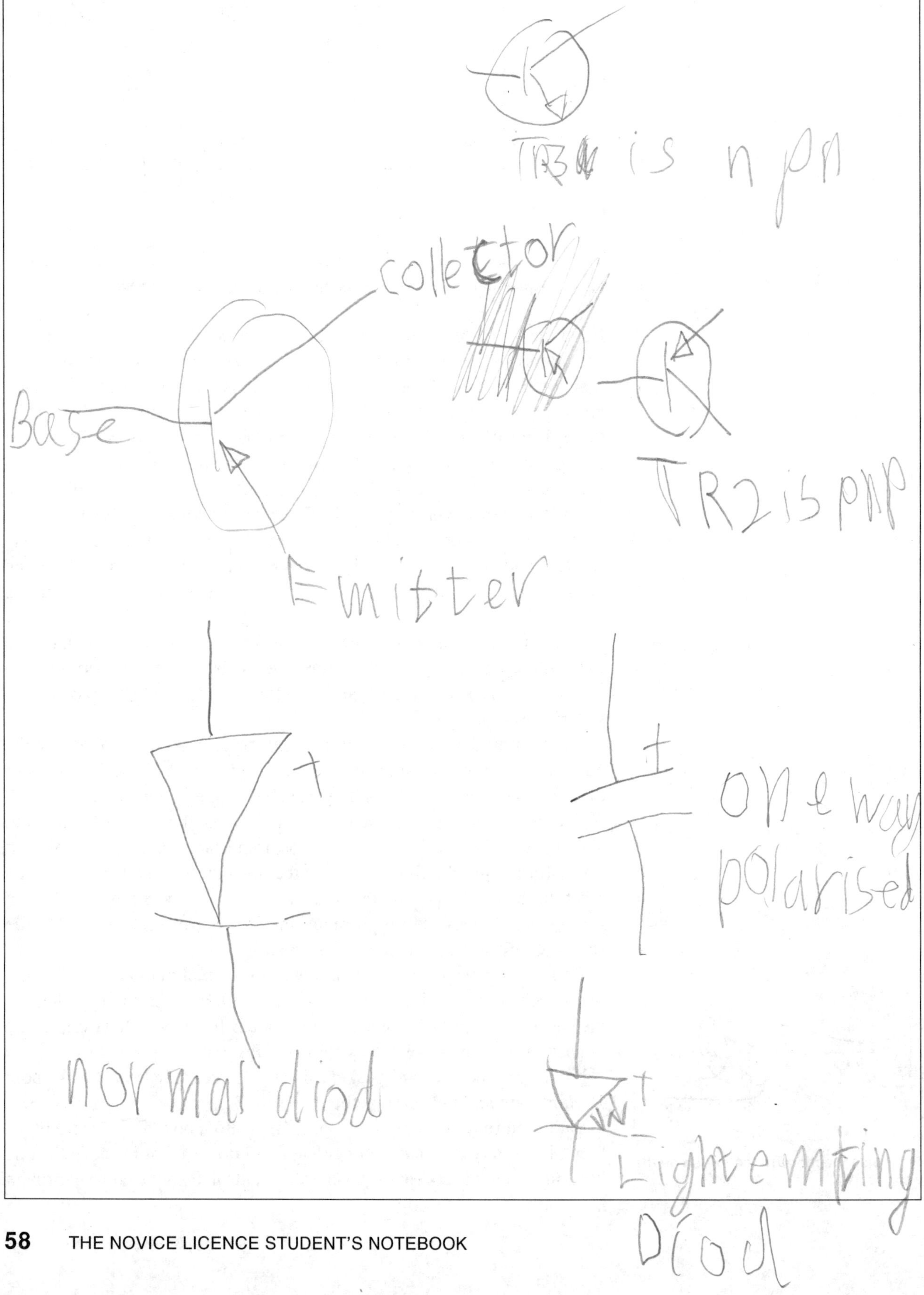
is npn
collector
Base
Emitter
TR2 is pnp
normal diod
one way
polarised
Light emting
Diod

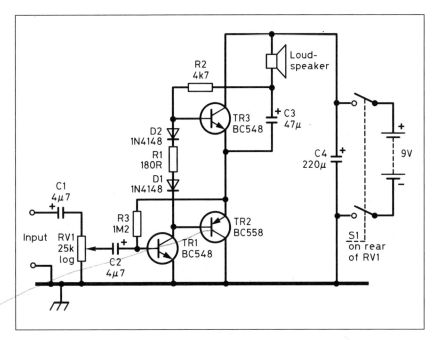

Fig 11. The amplifier's circuit diagram

the hole and tighten the nut – the PCB will be supported by the control. The prototype was not put into a box but mounted on an odd piece of aluminium which was screwed to a piece of wood to form a right-angle. The loudspeaker was fitted to the aluminium by means of two small pieces of aluminium with 3mm holes drilled in them. These are used as clamps, just holding the edge of the speaker. To allow the sound to get out, a few holes will need to be drilled in the panel (before the loudspeaker is fitted!) within a circle of about 40mm.

The signal to be amplified can be connected to the two input 'pads' by means of two short pieces of wire but if the connection needs to be a long one, then screened wires must be used.

If you decide to use a different loudspeaker make sure that it is at least 35 ohms impedance – anything lower will probably damage the two output transistors and even if it doesn't, it will cause your battery to run down rather quickly. If you have already built a crystal radio receiver or perhaps the MF receiver described in Part 2, this little amplifier will make the output sound really great.

Just for interest the circuit diagram of the amplifier is shown in Fig 11.

Notes

☞ Syllabus section 2(b)

The object is to see for yourself some of the effects of changing components in a simple circuit. The things you need are:

- Your Test Set No 1
- Two AA (HP7) cells
- A dummy cell
- A soldering iron and some solder

Put an AA cell in one battery holder and the dummy cell in the other.

Join X and Y – the bulb should light. It's not very bright. Gently touch the bulb, the 8.2 ohm resistor and the battery with your finger. Do you notice anything? What?

What can we do to make the bulb light brighter? There are three things – have you thought of them all ?

First disconnect X and Y. De-solder the link from A and make sure that it is long enough to touch either A or B. Reconnect X and Y – touch the link to A and then to B. Report the result.

Disconnect X and Y and solder the link back on to A. Now remove the dummy cell from the battery box and put a second AA cell in its place so that the two cells are facing the same way. Join X to Y and describe the result. What happens?

Disconnect X and Y. Turn one of the cells so that they are facing in opposite directions and connect X to Y. What happens? Touch the bulb, the 8.2 ohm resistor and the battery and describe what you feel.

Disconnect X and Y, move the link to B and reconnect X and Y. Again touch the bulb, the 2.2 ohm resistor (which is now in circuit) and the battery. List all the results.

Remove both AA cells and replace them in the reverse position – touch X and Y together. What happens?

For the last test get a short piece of wire and temporarily connect D to C. What happens now?

Now make a list of the things you have done, the results obtained and then describe what you have discovered from the tests.

Notes

☞ Syllabus sections 3/3.11

When the switch/es on a multimeter is/are set to measure resistance, a number of important things happen. Let us look at the analogue meter first but, before we try anything, it is important to know that all ohmmeters measure resistance by actually measuring the amount of current flowing through the resistor under test. This current is often provided by a battery inside the meter but in some mains-operated meters it will be provided by the power supply. In most analogue meters this causes the scale to be the other way round; in other words, very low resistance (zero ohms) is at the right-hand end of the scale and very high resistance (infinity) is at the left-hand end. This is because the battery will cause more current to flow through a low resistance and less to flow through a high one. When the leads of the meter are shorted together the pointer moves to the right-hand end of the scale (zero ohms).

Switch the meter to read ohms; check that the red lead is plugged into the positive socket and the black lead into the negative socket. Connect the two leads together – the pointer will move across the scale and stop somewhere near the '0' in the ohms scale. As the battery gets older its voltage changes; to allow for this there is a control marked 'zero ohms'. Adjust this until the pointer is exactly on the '0' of the scale. The meter is now ready to measure ohms.

Use some of the resistors you had for the colour code exercise. Connect the meter leads to the wires of each one in turn and read the values from the ohms scale. Now change the leads over. Does the resistance change? Read the value indicated by the colour code and compare with the value given by the meter. If the pointer stays at the high-resistance end of the scale, change the meter to a higher range. Usually the range switch will be marked '×100' so that a reading of 500 ohms actually would be 50,000 ohms (50kΩ). Yet another position of the range switch may be marked '×10,000'. On this range a reading of 75 ohms actually would be 750,000 ohms (750kΩ). On some meters there will be a zero ohms control for each range so when you change ranges connect the leads together to see if the pointer moves to the '0' on the ohms scale.

Now measure the 'resistance' of some different components. When checking other components, always reverse the leads of the meter and test again. You can try any components except batteries but try especially three different types of capacitor: (a) polyester, about 100nF (nanofarad), (b) electrolytic, about 100μF (microfarad) and (c) tantalum, about 4.7μF. Report exactly what happens in each test. Now check a diode; any type will do but you will probably have a 1N4148 (a general purpose switching type). Again report your results.

Use the digital meter to test the same group of components. Once again you will find it much easier to use.

Set the digital meter to read volts and use it to test the voltage between the leads of the analogue meter while it is switched to ohms. Take special note of the polarity of the indicated voltage, that is, which terminal of the analogue meter appears to be positive.

Repeat the same test but this time use the analogue meter as a voltmeter to measure the voltage of the digital meter when it is switched to ohms. Re-check your results to be sure!

Finally, before you leave the meter, switch it to a high-voltage DC range or, better still, to the OFF position if there is one.

Don't forget . . .

When you use an analogue ohmmeter to check an electrolytic or tantalum capacitor it is most important to remember that the black lead (normally negative) of the meter actually becomes positive and must be connected to these types of capacitor correctly. Similarly a diode will appear to work incorrectly if you forget this fact.

Notes

☞ *Syllabus section 8*

By this time you should be able to solder quite well, so no instructions other than the diagram in Fig 12 will be given for the construction of this unit. If you have any doubts, use the same techniques as you did for Test Set No 1. The real purpose of the exercise is the production of another circuit which will be used for practice in measurements later on.

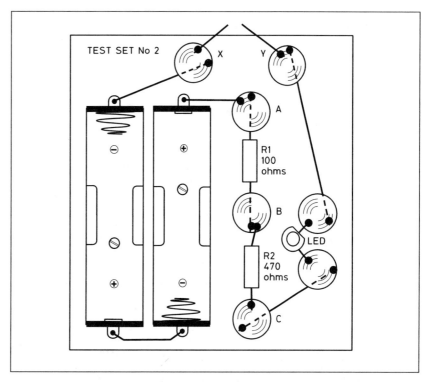

TEST SET No 2

R1
100
ohms

R2
470
ohms

LED

Fig 12. The layout of Test Set No 2. Note the 'flat' on the LED

When the circuit is complete, check it carefully, then put an AA cell in each battery box. You must decide which way they are to be put in! Connect X to Y and report what happens. If the LED (light emitting diode) does not light, look at your batteries – have you put them in correctly? Try turning one around. Try turning both of them around. If you still have trouble, see if your instructor can help you. Write a short statement in your report to say what you think is happening.

Short-circuit (connect a piece of wire across) the 470 ohm resistor and say what happens.

Again label the pins, using the same letters as those used in the diagram. Write your name on a sticky label and fix it to the under-side of the base of the unit. You will need Test Set No 2 for more exercises later.

Remember that this and all of your exercises will be marked and the marks will affect your final assessment, so always try to make your work very good – better than the model made by your instructor if you can!

Notes

The Radio Society of Great Britain, in common with many other radio societies in the world, has organised a system which allows members' QSL cards to be distributed to clubs and individuals throughout the world. Sending cards through the ordinary mail can be very expensive – the QSL Bureau will save members a considerable sum of money and will reduce the time spent on sending their cards.

The scheme works something like this. A member wishing to send cards first sorts them into countries and sends a complete package to the QSL Manager at RSGB Headquarters. The cards for a particular country are put with others for the same country and sent in bulk to the QSL Manager of that country. He will then sort the cards into regions and send them to a sub-manager in that region, and they will then be sent to the amateur to whom they are addressed.

To allow this to work, it is the responsibility of each member to provide his/her sub-manager with stamped and addressed envelopes and to indicate how many cards, up to the maximum, they wish to go into the envelopes. When this number of cards have been collected by the sub-manager the envelopes are posted.

If you are not a member of the RSGB, or if you wish to get a reply quickly, you can send your cards by ordinary mail. If you want to be fairly certain that you will get a card in return you must include an International Reply Coupon (IRC) which is rather expensive – to send just one card by post other than airmail, together with one IRC, will probably cost about £1.50 and then the reply will come by surface mail and can take quite a long time. Of course, if the card is being sent to an amateur in the UK, a stamped and addressed envelope is all that is necessary.

To send a card through the bureau you only need to know the callsign of the station but if you are going to post direct you will need to know the address. Usually amateurs are not very willing to give their address over the air, so you will need to be able to look up the callsign in either the *RSGB Amateur Radio Call Book* (callsigns starting with 'G', '2' or 'EI') or in an international call book for all other callsigns. The international books *Volume 1, The Callbook – United States Listings* and *Volume 2, The Callbook – Foreign Listings* cover the whole world. They are rather expensive but many clubs and some public libraries will have copies which you could probably look at.

You may often hear a strange callsign and will want to know the country to which it belongs. You will find this information, together with lots of other very useful things, in the front of the *RSGB Amateur Radio Call Book* or in the *Amateur Radio Operating Manual*, also obtainable from the RSGB. It would be a most useful book for you to own.

Notes

☞ *Syllabus sections 3/3.9*

When an analogue meter is switched to measure current it is *very* important that you remember that the meter is most vulnerable in this state – that is just a posh way of saying it is very easily damaged, so please be very careful and ask if in any doubt at all.

For the reason stated above you should never leave the meter switched to a current range – switch it back to a DC voltage range *even if you think you are going to use it as an ammeter in just a few minutes time.*

To measure current, the meter must be connected in series with the circuit so that the current flows through the meter – it is almost a short-circuit, so it won't alter the current flowing. The following test will help.

Use Test Set No 1. Fit one AA cell and a dummy cell. Connect X to Y to check that the bulb (lamp) lights. Now disconnect X and Y. Set the meter range switch to 1A DC and connect the meter leads to X and Y. Because the meter is almost a short-circuit the bulb will light as brightly as before but the current is now flowing through the meter as well as the bulb and the rest of the circuit.

Try to read the scale. Disconnect and reconnect one lead of the meter. Can you see any movement of the pointer? If yes, which way? If the pointer tries to move backwards (to the left), reverse the meter leads and try again. Move the range switch to 100mA DC – you should now be able to see enough movement to allow you to read the scale. You should record the current (through the bulb) as so many milliamps or, if you wish, divide the reading by 1000 and express the current as 0.?? amps.

Disconnect the meter. Short X to Y and the bulb lights again. Now, with the meter still switched to the 100mA DC range, connect it between the terminals of the bulb-holder – just as if you were measuring voltage. What happens? Do you know why? If not, ask your instructor. Connect the meter between C and D. What happens? Why? Again, if in any doubt ask your instructor. You have been using the ammeter as if it was a voltmeter; you have seen some of the effects of doing this and it could be much worse. In future we will avoid connecting an ammeter in that way!

If there is time replace the 'dummy' with a second AA cell and do some of the tests again.

Notes

Fit a 13A plug to three-core mains lead

This is one of the assessment exercises so you must keep trying until you get it right!

Three-core, flexible mains lead is normally round with an outer sheath of rubber or PVC. Inside there are three cores – multi-stranded wires, each insulated with rubber or PVC which are coloured brown (live), blue (neutral) and yellow/green stripes (earth). Great care must be taken with any conductors which are to be connected to the mains supply, so if any of the following steps go wrong, cut off the end of the cable and start again.

Step 1

Remove about 40mm of the outer sheath of the cable. The best way to do this is with the help of a sharp modelling knife. Run the knife around the cable about 40mm from one end. The knife should cut into the sheath but not deep enough to touch the insulation of the cores underneath. Now make a light cut from the end of the cable almost to the first cut; again do not cut so deep as to damage the cores. With a pair of side cutters, grip the edge of the outer at the end of the cable and peel it back to the circular cut then, using fingers, pull the outer sheath off. Bend the cable while turning it to examine the insulation of the cores – there must be no cuts visible at the point where they emerge from the outer sheath. If there are, cut the ends off and start again.

Step 2

Remove the cover from the plug. Inside, where the cable goes into the plug, there is a strip (usually of fibre) held in position by two screws. This is the cord grip. A glance at Fig 13 should make this clear. Undo the two screws until it is possible to pass the cable outer insulation through the space between the cord grip and the body of the plug. The outer should just be visible on the inside of the grip. Do not tighten the screws yet.

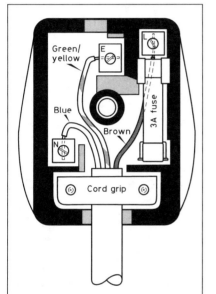

Fig 13. The connections to a typical 13A plug top

Step 3

Again referring to the diagram, find the terminal marked 'E' ('earth') and lay the yellow/green wire through the plug. Cut the wire so that it is a bit longer than the far side of the terminal. Now find the terminal marked 'N' ('neutral') and lay the blue wire in the body of the plug so that it reaches just beyond the terminal and cut it. Lastly find the terminal marked 'L' ('live') and lay the brown wire in the plug, then cut it so that it is just longer than the terminal. The three wires are now the correct length for fitting. Remove the cable from the plug and remove about 5mm of insulation from each core. Use the same method as before – cut lightly into the covering, not deep enough to cut into the wires beneath, then with the help of the side cutters pull the covering off the wire. There must be no broken strands which could produce a short-circuit and probably a 'blown' fuse.

Step 4

When you are sure all is well, examine the cover of the plug. Sometimes the cable has to be passed through a hole and, if this is the case with your plug, now is the time to do so. The three prepared ends are put into the hole from the outside towards the inside of the plug. Do it that way even though you could put the other end of the cable through in the other direction, but remember that in practice the cable would disappear into a piece of equipment and only the prepared end would be free. Push the

Notes

ends through the cord grip until the outer is just visible on the inside, then tighten the two screws to hold the cable firm. Twist the wires of the green/yellow core tightly together, slacken the terminal screw marked 'E' and push the wires into the hole. Tighten the screw very firmly. Repeat the operation with the blue core into the terminal marked 'N' and finally the brown core into the terminal marked 'L'.

Step 5

Look at the fuse and read the 'rating'. In practice you should fit a fuse which has a rating just enough to run the equipment. A 13A fuse should never be used with a 25W soldering iron but a 3A fuse would not last long if used with a 2kW electric fire! Finally fit the cover to the plug and show it to your instructor.

Later you may find that a piece of equipment is fitted with a flexible mains lead having only two cores. These will normally be coloured brown (live) and blue (neutral) and connected to the pins marked 'L' and 'N' respectively. When fitting a 13A plug to this type of cable, the earth pin remains unconnected. *The pin must not be removed from the plug.*

Notes

Fit a PL259 plug to coaxial cable

Many failures in amateur equipment are caused by incorrectly fitting plugs on to coaxial cable, so it is important to learn to make these connections correctly. Also, the exercise is one of the assessment tests, so take great care.

Coaxial cable is used widely for connecting the aerial to the transceiver and to connect other pieces of apparatus such as the ATU, the power output meter and dummy load. It is important that these leads do not themselves act as aerials and radiate signals. Signals which are radiated from cables and directly from apparatus are much more likely to cause interference with television (TVI) and other domestic apparatus. The cables used are normally screened to reduce the possibility of this happening.

Examine the end of a piece of coaxial cable. You will see that it consists of a central conductor (which may be a solid wire or it may be stranded). This is insulated by means of PVC and over the top is a woven wire braid. There is then an outer covering of insulating material which once again is usually PVC. The central conductor is often referred to as the 'inner' and the braided wire as the 'outer'.

Now look at Fig 14. Prepare the end of the coaxial cable by removing about 30 to 40mm of the outer covering, using the same techniques as for the mains cable. It is most important that you do not cut into the outer braid as the fine wires will break off and probably cause a short-circuit at some later time. If you cut carefully around the outer insulation of the cable and at the same time bend the cable, the insulation will break at the cut before it is deep enough to cut into the braid. Another light cut down the length of the cable from the first cut to the end will enable the outer covering to be peeled off. Using some pointed instrument (a pencil will do), tease out the strands of the braid right back to the outer covering.

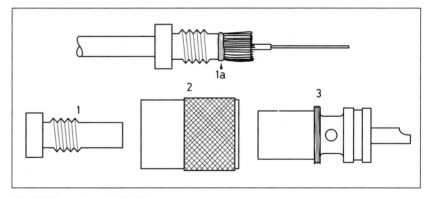

Fig 14. The parts of a PL259 plug

Prepare the reducer (1) by cleaning the plain end just beyond the thread; use wire wool or a knife to scrape the surface clean. Hold the reducer in a vice or a clothes peg and tin the surface you have just cleaned using a fairly large bit in the soldering iron. Allow to cool. Thread the reducer over the end of the cable. Cut the ends of the braid about 25mm from the end of the outer and fold them back over the reducer. Try to spread them evenly around the plain end of the reducer as shown in Fig 14 (1a).

Solder the ends of the braid to the reducer, keeping the wires close to the surface of the reducer and soldering only about 3mm. Refer again to (1a) in the diagram. You will find this operation easier if you wind a few turns of cotton around the braid wires to hold them in position while they are soldered and then remove the cotton when the part is cool.

Notes

Before going on to the next stage, pass the ferrule (2) over the reducer, keeping it the same way round as in the diagram. This is most important as it will not pass over the body of the plug after it is fitted.

Next cut the insulation around the inner about 3mm from the braid – again refer to Fig 14 (1a). Be very careful not to cut into the inner conductor. Use the same technique as before, a light cut and bend at the same time.

When the insulation is cut all the way round, remove it by pulling carefully with side cutters. If the inner conductor is stranded, twist the wires together to stop them doubling back during the next part of the operation.

Guide the inner through the body of the plug (3) until it appears at the pin end of the plug. When you are sure all of the wires of the inner have come through, screw the reducer into the plug and tighten with pliers. Solder the inner where it comes out of the pin of the plug, being careful not to use too much solder which could increase the size of the pin – if this happens, the plug will not fit into the socket. Finally screw the ferrule over the body of the plug until it becomes free. The connector is complete!

Notes

Like the PL259 plug, these smaller plugs are often used to connect the various parts of an amateur system together. Because they are smaller, these plugs are a bit fiddly to fit.

There are two different types in common use – both are shown in the diagrams and differ in one respect only: the way in which the outer braid is connected to the body of the plug.

Examine the parts and compare with the diagram (Fig 15). The difference is in the part labelled '4a' and '4b'.

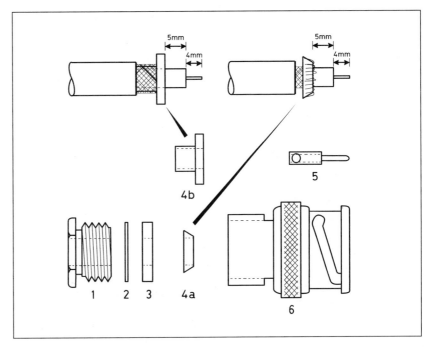

Fig 15. The parts of a BNC plug

Start by removing about 30mm of the outer cover, again being very careful not to damage the braid. Note that 30mm is more than really necessary but beginners usually find it easier to handle the longer length. Thread on the parts starting with the nut (1), put it on in the same direction as shown in the diagram then follow with the plain metal washer (2), the rubber washer (3) and the cone (4a). This last one must go on in the direction in which it is drawn and may have to be coaxed over the ends of the braid. Push it up until it is touching the outer covering.

As with the PL259, use a pointed instrument to tease out the wires of the braid all of the way back to the conical washer. Trim the wires of the braid to about 10mm – a pair of scissors will be useful to do this. Bend the ends of the braid back over the cone as shown in the diagram. Now remove the insulation from the inner so as to leave just 5mm sticking out from the cone, then cut the inner to leave 4mm of inner sticking out from the insulation. These last two dimensions are most important as they determine the position of the pin in the finished plug.

Fit the pin (5) to the inner and carefully solder through the small hole in the pin. Heat the pin and apply the solder (22swg) to the hole. *Be very mean with the solder* as too much will stop the parts fitting together. Give the pin a pull to make sure that it is soldered.

Push the cable, together with the pin and the cone, into the body of the plug (6), watching that the braid remains wrapped around the cone. The pin will only just pass through the hole in the insulation inside the plug

Notes

body so you will have to apply a little patience to make it fit. Various makes of plug use different methods of providing the insulation. Sometimes it is already fitted inside the plug body or it may be in the form of a small white sleeve which must be pushed into the body before the pin is fitted. Another make supplies the insulation in two halves so that it can be fitted around the pin before the assembly is pushed into the body. Your instructor will help you decide which type of plug you have if there is any doubt. Check that the end of the pin is about level with the edge of the body of the plug. When you are satisfied, gently push the rubber washer into the body and follow it with the plain metal washer. It should now be possible to screw the nut into the body and tighten it up. As you do so the braid will be pushed hard up against the cone and the body of the plug, making the connection between braid and plug body.

If instead of a cone you have a part like the one shown at 4b, the process is slightly different and probably a bit easier. Fit nut, plain washer and rubber washer as before. Then remove the outer and the braid so as to leave about 30mm of inner sticking out. Outer and braid are now level. Thread the ferrule (4b), thin end first over the inner, and push it up towards the braid. Continue so that the thin end of the ferrule goes *underneath* the braid and the thick end of the ferrule is touching the braid. Trim the inner and the inner insulation to the same dimensions as before, fit the pin and solder. Now assemble the plug as for the other type. This time the rubber washer will squeeze the braid on to the ferrule and so make a good connection to the body of the plug.

Notes

Block diagrams are often used to explain the function of a receiver, a transmitter or almost any electronic device without giving much detail. At this stage we only need to know what the bits inside the box are expected to do. We do not need to know how they do it, or what is inside the box.

Tuned radio frequency receiver

Fig 16 shows a type of radio known as a 'tuned radio frequency' (TRF) receiver. The aerial will provide very small voltages from a number of different transmitters. The voltage may be less than one microvolt, that is, less than a millionth part of a volt. These very weak signals are applied to the box marked 'RF amplifier'. This will make a big copy of the small signal (just as a slide projector makes a big copy of the small picture on the film) and will pass it on to the next box which is marked 'Detector'.

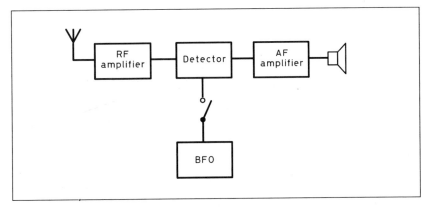

Fig 16. A tuned radio frequency (TRF) receiver

The job of the detector is to sort out the audio frequency that has been mixed with the RF at the transmitter and to pass it on to the next box, which is an 'AF amplifier'. This will make a big copy of the audio frequency signal from the demodulator and will provide enough power to operate either headphones or a loudspeaker. A loudspeaker will require more power than headphones but the principle of the amplifier is not altered.

The remaining box contains the letters 'BFO' which are short for 'beat frequency oscillator'. This is included to indicate the principle often used to make an unmodulated signal (continuous wave) audible. The method is rarely used in a TRF receiver but is common in other types.

If Morse code signals are being received there will normally be no audio frequency in the signal. The Morse key simply switches the RF from the transmitter on and off in the form of dots and dashes but we cannot hear the effect of RF. To make it audible we use an effect known as 'heterodyning'. This occurs whenever signals of different frequencies are mixed together. New frequencies are produced and one of these is equal to the difference between the two originals. For example, if an RF signal having a frequency of 3.55MHz is being received, it would not be audible. If the BFO is switched on and tuned to a frequency just 1kHz higher than 3.55 MHz, that is, to a frequency of 3.551MHz, a new frequency of 1kHz will be added to the signal applied to the detector which will separate it from the mixture and send an audio signal to the AF amplifier.

TRF receiver with reaction

Fig 17 shows a different type of TRF receiver. In this circuit the demodulator can be made into an oscillator by adjusting a control called

DEMODULATOR AND DETECTOR ARE THE SAME THING.

Notes

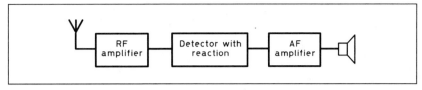

Fig 17. A TRF receiver with reaction

the 'reaction control'. If this is tuned to a frequency a little different to the received signal, the heterodyne effect takes place as before and the signal can be heard.

Superheterodyne receiver

Fig 18 shows a very simple form of another type of receiver which is called a 'superheterodyne' ('superhet' for short). This is the most common form of receiver in normal use. It has many advantages and some of these will be mentioned later.

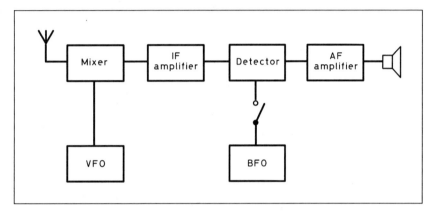

Fig 18. A superheterodyne receiver

The general idea is to change the frequency of all the required incoming signals to a new, fixed frequency. This allows the 'RF' amplifiers which follow to be tuned to that frequency, and no adjustment is necessary when different stations or different bands are used. To avoid confusion the new RF is called 'the intermediate frequency' (IF).

In Fig 18 the blocks 'IF Amp', Detector, 'AF Amp' and 'BFO' all behave as the four blocks in Fig 16, except that the working frequency is fixed. The mixer and the VFO (variable frequency oscillator) make use of the heterodyne principle to change the incoming signal to the new intermediate frequency (IF). In practice, in order to change from one station to another only the frequency of the VFO needs to be altered. This makes both operation and design very much simpler!

Don't worry too much about this, but do get to know the names and position of each block in the diagrams.

Notes

Demonstrate Ohm's Law – Test Set No 2

☞ *Syllabus section 2(b)*

Insert two AA cells into the battery boxes of your Test Set No 2. Short-circuit terminals X and Y. If the LED doesn't light look at the circuit carefully to see if you can spot anything wrong. How are the batteries arranged? Correct any error and try again.

When you have made the LED light, disconnect X and Y and measure the current flowing in the circuit by connecting an ammeter between X and Y. The meter range should be set to 100mA DC. If the current is less than 10mA, you can now move the range switch to the 10mA range. Note – this is the correct way of doing things if you do not know the amount of current flowing. You will have discovered that in a series circuit like this (where there is only one way for the current to go) the value of the current will be the same in all parts of the circuit, so the ammeter connected between X and Y will measure the current through all the elements of the circuit.

Now measure the voltage across R1, the 100 ohm resistor, and divide the value you get by 100. Compare the result with the measured value of current. Do the same thing again but this time measure the voltage across R2, the 470 ohm resistor, and divide the value by 470. In both cases you should have a result which is very close to the measured value of the current. This is not surprising because:

current (I) = voltage (V) divided by resistance (R)

Note that in both these tests you have used the voltage across a resistor and the value of *that* resistor. Also, the current given by the little calculation comes out as a fraction because the current in both cases is less than 1 ampere. Multiply the value by 1000 and the result will be the current in milliamperes (mA) which is the way the meter will display the value of the current.

This time connect the voltmeter between A and C and divide the value you get by the current which you measured earlier. The current must be in amperes (divide the meter reading by 1000). Your result should be very close to 570 ohms. Can you guess why? Resistance can be 'measured' by taking the voltage across a component and dividing by the current through that component.

$$R = \frac{V}{I}$$

Measure the voltage across the LED (connect the voltmeter between Y and C), then divide the value by the current (in amperes). The result is the resistance of the LED (in ohms). You should have realised by this time that when the voltage is to be measured, X and Y must be connected together to complete the circuit. Now remove the batteries and measure the resistance of the LED using the meter as an ohmmeter. Depending on the way you connect the meter, you will get two values for resistance but neither of them will be the value obtained by the tests in the previous paragraph. This is because the LED is a semiconductor and is non-linear, which means that the resistance will change with the value of current flowing through it. Which is the correct value to use – the calculated one or the measured one? Check with your instructor.

If the meter would not measure current – and some of the cheaper ones don't – we can find the value of the current by measuring the voltage across a resistor and dividing the result by the value of *that* resistor.

$$I = \frac{V}{R}$$

COVER THE ONE YOU WANT.

NOTE

← THUMB

MEANS $V = I \times R$

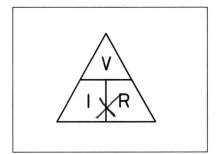

Fig 19. An easy way to remember three different formulae

Sometimes we may be unable to measure the value of a voltage, and in that case the value can be found by multiplying the current through the circuit by the resistance of the circuit.

$$V = I \times R$$

Now you have three simple formulae:

$$R = \frac{V}{I}$$

$$I = \frac{V}{R}$$

$$V = I \times R$$

To make it easier to remember them you may like to use the triangle shown in Fig 19.

Cover the one you want (*I*, *V* or *R*) and the value is indicated by the remaining letters.

Notes

☞ *Syllabus section 3.12*

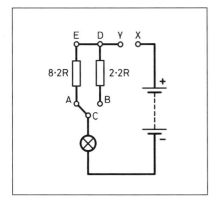

Fig 20. The circuit diagram of Test Set No 1

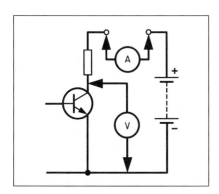

Fig 21. How the DC input power to a transistor is measured

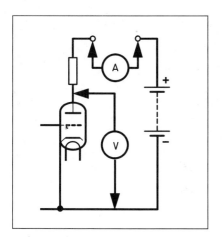

Fig 22. How the DC input power to a valve is measured

In this exercise you are going to make use of the multimeter to find the power of a circuit. What is 'power'? It is the rate at which energy is converted from one form to another. Power is measured in 'watts'.

When an electric fire is said to have a power of 1000 watts (1 kilowatt), it is an indication of how fast the fire can convert electrical energy into heat energy. There are meters which will measure power directly but you are going to use a multimeter.

First, here is a new way of drawing electrical circuits. The circuit diagram in Fig 20 is exactly the same circuit as the one in Test Set No 1 but the 'pictures' do not necessarily look like the actual component. It is easier and quicker to draw, and easier to read once you get used to the symbols. The letters which would not normally appear have been added so that it should be easier for you to compare this new diagram with the layout in Test Set No 1.

In the circuit the battery, the source of electrical energy, is causing current to flow through the lamp (bulb) which converts the electrical energy into light energy and into another energy form – can you say what it is?

Now let's try to discover the power of the lamp – the 'input power'. First measure the current through the lamp. Where do we connect the meter? Remember that to measure current, the meter must close a break in the circuit so we could connect the ammeter between X and Y. Try it. You may feel that this will not be the current flowing through the lamp, so try a few other measurements. Remove the link between A and C, short X and Y then connect the ammeter between A and C – perhaps that will indicate the current through the lamp. Just to be sure, replace the link A to C and disconnect the link between the two battery boxes, then connect the ammeter so that it acts as the link. Compare the three current measurements and then try to make a rule about the current in this type of circuit (known as a 'series circuit'). Check with your instructor.

Switch the meter back to a DC voltmeter and measure the voltage between the terminals of the lamp-holder.

Calculate the power of the lamp by multiplying the current through it by the voltage across it.

Power in watts (W) = Current in amps (I) × Voltage in volts (V)

In other words:

$$W = I \times V$$

The value you get is the input power to the lamp – the output of the bulb is light which would be measured in a different way and need not concern us at this moment.

Disconnect X and Y, remove the link A to C and connect it B to C. Now measure the input power to the lamp!

Don't forget to put the test set back to its original form.

The method you have just used can be applied to find the input power to any resistor or device that works as a resistor. The two circuits shown in Figs 21 and 22 look different but the input power can be measured in exactly the same way. In each case the meters would be connected as shown.

Notes

Direct current can be compared to marbles running down a ramp, something like a helter-skelter where the marbles will run from the high point (positive) to the low point (negative). The battery can be thought of as a device which takes the marbles up to the top again and so we will get a continuous stream rolling down the slope.

Instead of the simple ramp, imagine a see-saw with sides on the seat and marbles as before. If the see-saw is slowly rocked in the normal way, the marbles will roll down in one direction and then reverse when the platform moves to the other position. Again think of the high end as being positive and the low one negative. If we continually reverse the connections to a battery, current would flow in similar manner. An alternator is a generator of voltage which does change its polarity at a constant rate so that if a resistor is connected across the terminals of the alternator, current will flow first in one direction and then in the other – the current is called 'alternating current' (AC). This is the nature of current that flows through any device connected to the household electricity supply. If the current starts at nothing (zero), it increases to a maximum value in one direction, decreases back to zero and then commences to increase in the opposite direction until it reaches the same maximum value as before, when it grows less and again becomes zero. The complete sequence just described is called a 'cycle' and the number of times that the current does this in one second is called the 'frequency'. This is measured in cycles per second or 'hertz'. One cycle/second is one hertz. The frequency of the domestic supply is 50 hertz (50Hz). Fig 23 illustrates these effects.

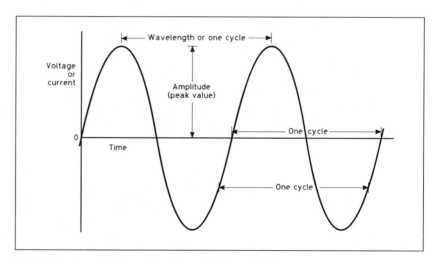

Fig 23. An alternating waveform illustrated by a graph

When radio frequencies are being considered, the dimension between the tops of the wave marked 'one cycle' is sometimes referred to as the 'wavelength'. There is more about this in Worksheet 28 – 'The spectrum'.

The current which flows in the radio-frequency portions of a radio transmitter will have a frequency very much higher than 50Hz – radio frequencies start at about 20,000Hz (20kHz) and continue up to about 200,000,000,000Hz (200 gigahertz – 200GHz). You will have come across frequencies while you have been operating the radio and will know that the lowest frequency band used by radio amateurs is 1.8 to 2.0MHz. Try to find out the frequencies used in all the other amateur bands.

Try to obtain some back copies of *D-i-Y Radio*, Volume 2, parts 1, 2 and 3 – you will find some very useful articles entitled 'All waves great and small' by Mike Dixon, G3PFR.

Notes

☞ *Syllabus section 3(a)*

A simple aerial system such as a long wire will receive weak signals from many transmitters and it is necessary to pick out the station required. The 'tuned circuit' is often used for this purpose.

You have already seen an experiment where one pendulum is caused to 'oscillate' by absorbing energy from another pendulum with the same frequency. A circuit consisting of a coil and capacitor will have a natural frequency and will respond to energy from a transmitter with an output at that frequency. It will not respond to energy at other frequencies and so the unwanted stations do not make the circuit oscillate. When the tuning control of a receiver is moved, it is probably altering the frequency of a tuned circuit. In a medium-wave or long-wave receiver a coil is often wound on the ferrite rod used as an aerial and the tuning control is altering the capacitance of a 'variable capacitor'. Fig 24 shows two tuned circuits which sometimes are used for the purpose just described.

The pendulum in a clock can be made to swing by taking energy from a spring or a system of weights or a battery. A tuned circuit can be made to oscillate by taking energy from a battery or other supply. It is just like a swing. Someone pushes the swing at just the right moment in time and as long as they keep doing so the swing continues to 'go'. A transistor or a valve can be used to release energy from a battery at exactly the right moment and keep the oscillation in the tuned circuit going. The complete device is called an 'oscillator'. An oscillator is the part of a transmitter which generates the frequency to be radiated.

The use of oscillators in transmitters is shown in Worksheet 26.

Many aerials used by amateurs are actually large tuned circuits. A dipole is an excellent example, and in your earlier experiments you found the frequency of an aerial by measuring the length of the rods and then carrying out a simple calculation. This should tell you that tuned circuits do not always look like those in the diagrams in Fig 24.

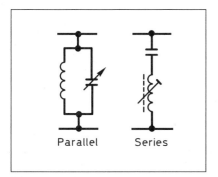

Fig 24. Two forms of tuned circuits: (a) parallel and (b) series

Notes

Much of the equipment in an amateur radio station needs some form of electrical energy for its operation. There are two main sources of this energy: batteries and the domestic mains supply.

Batteries

These may be dry, lead/acid or alkaline cells.

You will be familiar with dry cells as they often provide the energy for equipment such as personal tape players, portable radios, radio-controlled cars and other models, torches and many other devices. The main advantage of the dry cell is its portability and the disadvantage is the high cost. Dry batteries are a very expensive source of electricity (several hundred times as much as the energy from the domestic mains supply!) but we pay for the convenience.

Rechargeable cells are usually nickel/cadmium ('ni-cad' for short) and are obtainable in the same sizes as dry cells. They will therefore fit into the equipment with no modification. They can cost two to three times as much as the best-quality dry cells but, because they can be recharged, they can be used many times. The cost of charging is very small and they eventually work out much cheaper. Of course, it is necessary to buy a suitable charger but with care this will last almost for ever.

Lead/acid cells are generally used where high current is required and the radio amateur will normally use these only when operating equipment in motor vehicles. When radio equipment is fitted into a car, great care must be taken to ensure that the battery is protected against a short-circuit. A car battery is capable of providing a current of several hundred amperes and currents of this order will cause even thick wires to become red hot. Burns and fires are therefore a real hazard and the use of a fuse, close to the battery, is essential. Before making any connection to the car supply, the battery should be disconnected. Hacksaws and spanners have been known to become red hot by getting 'stuck' across the terminals of the battery.

The domestic mains supply

The great advantage of this source of energy is the very low cost but there are a number of disadvantages.

(a) Any device connected to the mains cannot be truly portable.
(b) Great care must be taken with the design to avoid electric shock.
(c) There is a big difference between the supply voltage (240V AC) and that required by amateur equipment – usually about 12V DC.

(b) and (c) together determine the design of the power supply. Metal boxes (correctly earthed), fuses in the AC input and the DC output lines and a means of indicating that the unit is switched on are mandatory. The last item is best provided by an indicator lamp.

The lower DC output voltage is normally obtained by the use of a transformer to reduce the voltage and a rectifier to change the AC into DC.

Notes

You have already seen block diagrams in Worksheet 20 and will now have an idea of their use. Your instructor will have talked about the following diagrams, so only the main points are included in this sheet.

A Morse code transmitter

See Fig 25.

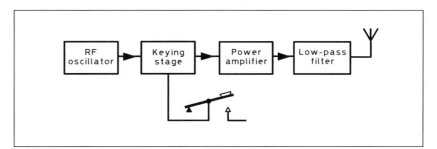

Fig 25. A Morse code transmitter

The RF oscillator

This block appears in all four diagrams and has the same function in each, that is to generate the radio frequency which is to be transmitted and will act as the 'carrier' of the 'data' such as Morse code, speech or television signal (video).

The keying stage

This is necessary in a Morse code transmitter as it is unwise to switch the oscillator on and off. It takes time for the oscillator to 'stabilise', that is to reach a constant frequency. It is also undesirable to switch the power amplifier as this tends to produce annoying clicks at the beginning and end of the dots and dashes. The keying stage is really an amplifier but is dealing with a fairly small signal.

The power output stage

This is another amplifier but is arranged to give sufficient power to feed into the aerial so that a good signal is radiated.

The low-pass filter

This is a most important device which will allow the carrier frequency to pass through but will stop higher frequencies such as twice that of the carrier (second harmonic) or higher frequencies still. If these were allowed to reach the aerial they would cause interference on frequencies which were not supposed to be in use.

An amplitude-modulated transmitter

See Fig 26. This is the first of three transmitters which can send out a modulated signal. The radio frequency signal is made to 'carry' other information – in this case speech.

A microphone converts the sound into electrical AF signals and these are amplified in an AF amplifier. The output of this amplifier is 'mixed' with the RF in the 'modulator stage' and once again the effect we have called 'heterodyning' takes place. New signals are generated which will have frequencies equal to the difference between the RF and the AF and also to the sum of the two. They are sometimes called 'sum and difference frequencies'. They are also given the name of 'sidebands'. You will hear this word quite a lot in future.

Notes

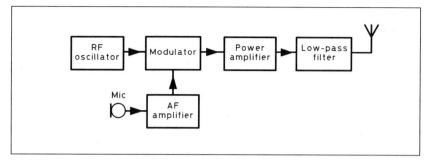

Fig 26. An amplitude-modulated transmitter

The carrier and the sidebands are amplified by the power amplifier and passed to the low-pass filter and aerial as before.

The arrangement shown is not the one used by transmitting stations where very good quality audio transfer is important. In broadcast transmitters the audio frequency would be applied to the power amplifier stage.

The great advantage of the system used in Fig 26 is that modulation can be carried out with low-power AF. To avoid too much distortion of the audio frequency the power amplifier must be of the type known as a 'linear amplifier'.

A frequency-modulated transmitter

In this transmitter (Fig 27) the AF is applied to the RF oscillator in such a way that it makes the frequency move up and down in time with the sounds at the microphone. Higher audio frequencies will cause the frequency of the carrier to increase and decrease rapidly while with lower AF, the carrier frequency will change more slowly. Loud sounds will make the frequency change a lot and quieter ones will cause only small changes.

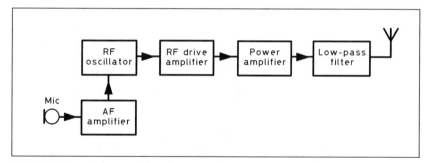

Fig 27. A frequency-modulated transmitter

When receiving this type of signal the receiver must have a different type of detector which is called a 'frequency discriminator'.

An SSB (single sideband) transmitter

This and the FM transmitter are probably the most common types in general use, so it is important to have some idea of the way in which it works. You will see that the diagram (Fig 28) is quite similar to the AM transmitter with only two changes.

The modulator is of a type known as a 'balanced mixer'. When the RF is mixed with the AF, the carrier is cancelled out in the circuit and only the sidebands appear at the output of the mixer. Only one sideband, either the sum of the frequencies (upper sideband) or the difference (lower sideband) is needed by the receiver in order for the AF to be demodulated. The next block, the 'sideband filter' will allow only one of the two sidebands to pass and so that a single sideband suppressed carrier – 'SSB' for short – is passed to the power amplifier, low-pass filter and the aerial.

Notes

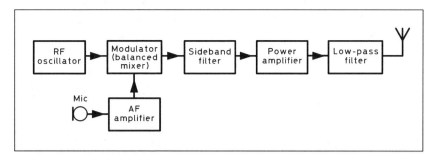

Fig 28. An SSB (single sideband) transmitter

Why go to all that trouble when an AM transmitter will do just as well? It is a question of space in the rather small bands allocated to amateur radio. An AM transmitter will need about 7kHz of space in the band, that is about 3.5kHz for each sideband. By getting rid of one sideband we can save half the space and so leave room for another transmitter. OK, but why get rid of the carrier? The carrier is only necessary at the receiver in order that the signals can be demodulated. The receiver can be made to provide its own carrier for that purpose and, as the transmitter does not have to send it out, the power amplifier has less work to do or can provide more power in the single sideband sent out.

Notes

Oscillators are so often used in radio equipment that a little more information about them will be helpful. Oscillators are just generators of either AF or RF and are designed so as to be stable (keep to a constant frequency) over long periods of time.

Generally we want the output from the oscillator to be a pure 'sine wave'. You have possibly met this shape if you have studied trigonometry but if not, the shape is shown in the worksheet about AC (Worksheet 23). In practice the waveforms are distorted (misshapen) by the circuits which generate them or by amplifiers etc. The change in shape introduces new frequencies which are called 'harmonics'.

If you have helped with the experiment with a length of elastic you will have seen that the strand can vibrate in a number of different ways, some of these are shown in Fig 29. The five pictures show simplified ways in which any 'string' could behave:

(a) shows the 'fundamental' which corresponds to a pure sine wave. If the frequency was within the audio range a very smooth, soft note would be produced.

(b) is the 'second harmonic'. The frequency is twice that of the fundamental or the wavelength is half. Again the waveform is a pure sine wave so we would hear a similar sound but one octave higher in the musical scale. In practice the string would vibrate in both ways at the same time and then we would hear a less musical note.

(c) is the third harmonic of (a),

(d) the fourth and

(e) the fifth.

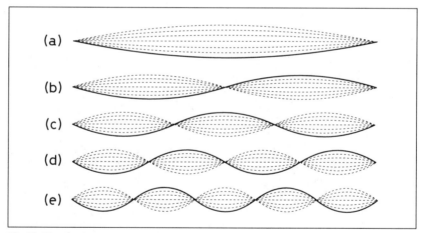

Fig 29. Harmonics

The strings in most stringed musical instruments will vibrate in a most complex way, combining at least all five waveforms at the same time and resulting in a rather strident type of note. If you are interested it would be a good idea to refer to Worksheet 6 and compare the ideas in the above paragraphs with the part about tone in the RST report.

If our transmitter radiates any harmonics it means that as well as the signal on the intended frequency, there will be a signal on frequencies which are double, three times, four times etc. Interference will now take place and, if it occurs on frequencies not allocated to amateur radio, it could have very serious results.

It is for this reason that our transmitters have a low-pass filter between the last amplifier and the aerial to allow the fundamental to pass but to greatly reduce any harmonics.

Notes

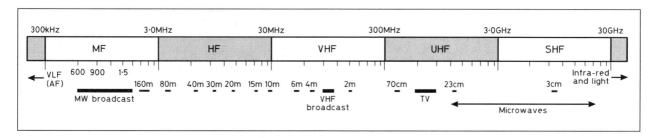

Fig 30. Showing where the various amateur bands fit into the spectrum

Fig 30 is drawn with a logarithmic scale so that the parts on the left-hand side are not too cramped. The first section, which is 300 to 3000kHz (0.3–3.0MHz), known as 'medium frequency' (MF), can contain just 300 stations each separated by 9kHz, which is in fact the space occupied by each of the stations in the medium- and long-wave bands of a domestic receiver. As you can see, all of the medium-wave band falls in this section.

Look at the second section, 3.0–30MHz, called 'high frequency' (HF). We find that there could be 3000 stations in the section with the same spacing as before and the section at the right-hand end which is 3000–30,000MHz (3–30GHz), called 'super high frequency' (SHF) or 'microwaves', could have 3,000,000 stations! It is for this reason that the logarithmic scale has been used. Along the bottom edge of each of the five rectangles you will see a number of graduation marks. These are to enable the frequency to be found at these points. Only some of the values have been put in as the space is rather limited. To find the rest, look at the frequency marked on the top left of a rectangle, double it to give the value at the first mark along the bottom of the section. Add the same number again to get the next value and so on.

The radio frequency spectrum extends beyond the range shown in the diagram but is not shown as there are no amateur radio frequencies available in those parts. Just for interest the long-wave band covers from 100–300kHz with Droitwich, the long-wave station, at 198kHz.

Some of the amateur bands are shown on the chart but because of the limited space a few of them have been missed out. At least all of those allocated to the Novice Licence are shown. When you are listening to a short-wave receiver and you discover the frequency of a station, try to find the place where it fits in the scale.

The velocity (speed) of an electromagnetic wave is constant at just about 300,000,000 metres per second. The wavelength in metres multiplied by the frequency in cycles per second (hertz) will always produce this figure. As the frequency increases the wavelength decreases. The simple formula:

$$v \text{ (velocity)} = \lambda \text{ (wavelength)} \times f \text{ (frequency)}$$

is a shorthand way of expressing the idea.

A triangle similar to the one in Worksheet 21 is shown in Fig 31 to enable one quantity to be calculated when the other two are known.

Now try this – if the frequency of a transmitter is 200kHz, what is the wavelength of the radiated wave? Check your answer with your instructor.

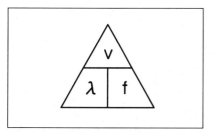

Fig 31. Another use of the formulae triangle

Notes

Test Set No 2 – with diodes and transistors

Fit your Test Set No 2 with batteries, and short-circuit pins X and Y to check that the LED lights. Remove one battery and then make the following alterations. Obtain a diode (1N4148 or similar) and solder the ends to pins X and Y. There is no need in this particular instance to make an excellent joint; we sometimes say 'tack the ends' which means just touch the ends with the hot iron and hold it there long enough to make the wires 'stick'.

Now replace the battery, making sure that it is the correct way round. Does the LED light? Remove the battery again, touch the wires of the diode with the hot iron to disconnect it and reconnect it the other way round. Put the battery back once more. Does the LED light? Think about the result and then try to give a reason why the effect you have produced occurs.

Take out one battery again – in fact from now on always do this before you solder to the circuit, as it will help to ensure that your diode etc will not be damaged. Disconnect the diode. Damp your finger and touch pins X and Y at the same time – don't worry, it won't hurt. Unless you have very special fingers the LED will not light.

You have already used transistors in your AF amplifier project and will be able to identify a BC108 or a BC548. Let's look again. The BC108 has a metal case with a little tab on the side. Look at the end, with the wires coming out towards you, and compare with the drawing in Fig 32. Starting at the tab, go around the wires in a clockwise direction. The first is called the 'emitter' (e), the second is the 'base' (b) and the third is the 'collector' (c). The BC548 has a plastic case with a flat part on it. Again look at the wires and compare with the second picture in Fig 32. Turn the transistor so that the flat part is at the bottom. The wire on the left is the emitter, the one in the middle is the base and the one on the right is the collector.

Solder the collector to pin Y on your test set and the emitter to pin X. Put the batteries in the boxes (in the correct direction). Unless you have chosen a 'dud' transistor the LED will not light. Again damp your finger and touch pin A and the third wire (base) of the transistor at the same time. The LED will light. *Do not touch the wire itself to pin A* – let your wet finger do the connecting! This is a simple way of using a transistor as a switch and also as an amplifier. The current flowing through your finger was not enough to light the LED in the first test but in this one the transistor behaves as a current amplifier.

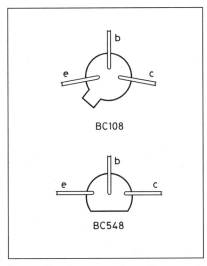

Fig 32. The pin connections of two types of transistor

Notes

Learning the Morse code

This worksheet is intended for those who hope to take the 5wpm Novice Morse test.

When learning Morse code, it may be found easier to organise the letters into some form of recognisable pattern and use 'mnemonics' (memory aids) to help in the learning of the code.

Each letter is made up of a series of short and long sounds – 'di's and 'dah's. When a 'di' is on its own or at the end of a combination, it is sounded as 'dit'; at all other times it is 'di'.

For example:

$$E = \cdot \quad \text{(dit)}$$
$$A = \cdot - \quad \text{(di-dah)}$$
$$R = \cdot - \cdot \quad \text{(di-dah-dit)}$$

Here is the full code for the letters:

A = ·−	H = ····	O = −−−	U = ··−
B = −···	I = ··	P = ·−−·	V = ···−
C = −·−·	J = ·−−−	Q = −−·−	W = ·−−
D = −··	K = −·−	R = ·−·	X = −··−
E = ·	L = ·−··	S = ···	Y = −·−−
F = ··−·	M = −−	T = −	Z = −−··
G = −−·	N = −·		

There are, however, many ways to learn Morse code and it is always best to choose the method which is most suited to your way of learning.

Now here is *one* way of learning it.

Too	T	−
Many	M	−−
Operators	O	−−−

Notice the triangle of dahs. If you remember "too many operators" goes with a triangle of dahs; you need only know T = dah to remember the rest.

Eric	E	·
Is	I	··
So	S	···
Happy	H	····

A triangle of dits along with your mnemonic "Eric is so happy" (or any which you can remember). If you know E = dit, then along with your little saying and the dit triangle, you will know these letters.

Are	A	·−
U (you)	U	··−
Voting	V	···−

A triangle of dits followed by dahs. Learn 'A', remember the mnemonic of your choice and the pattern, and you will be well on your way.

Aunty	A	·−
Wendy	W	·−−
Jogs	J	·−−−

You already know 'A', so remember your saying and the dah triangle with a preceding dit.

Now you have learnt nearly half the alphabet, so just for a change let's look at the numbers – they are so easy.

1 = ·−−−−			6 = −····
2 = ··−−−			7 = −−···
3 = ···−−			8 = −−−··
4 = ····−			9 = −−−−·
5 = ·····			0 = −−−−−

Notes

All the numbers are made up of five parts, whereas the biggest combination in the letters is four. Note that in the numbers one to five, the dits indicate the number: four is four dits, two is two dits and the remaining spaces are filled in by dahs.

When it comes to the numbers from six to zero, we have to store five in our minds, and add the dahs to that five. The remaining spaces are filled in by dits, eg 7 is − − · · · (five in the mind and two dahs = 7). The 0 (zero) you just have to learn!

There is one last triangle of dits and dahs to remember.

No	N	− ·
Dog	D	− · ·
Biscuits	B	− · · ·

A triangle of dits preceded by dahs. Learn 'N', remember the pattern and your mnemonic and there you are.

Now the learning pattern changes. We have groups of two letters with opposite patterns.

Queen	Q	− − · −
Fox	F	· · − ·

Learn 'Q' and remember the mnemonic and the fact that the patterns are opposite.

King	K	− · −
Rat	R	· − ·

The same principle applies.

Play	P	· − − ·
Xylophones	X	− · · −
Long	L	· − · ·
Year	Y	− · − −

Things are never perfect and so there are a few letters that don't fit into any pattern of learning code – these just have to be learnt on their own.

G	− − ·

This will be the first letter of many UK callsigns.

C	− · − ·
Q	− − · −

CQ is the call that amateurs put out over the air when asking for a contact.

Z	− − · ·

The last letter.

Although you don't have to learn Morse code off by heart to complete the course, this worksheet will be useful between the end of the course and the C&GLI examination, for those intending to take out a Class A Novice Licence.

Note that when sound signals are used, the length of a dash is three times as long as a dot. The space between the dots and dashes in a character is equal to one dot and the space between letters is equal to three dots (or one dash).

Notes

The format of the Novice Morse Test

The Novice 5wpm Morse Test is designed to prepare candidates for CW 'life on the air'. The test will comprise a receiving and sending test. The receiving test will normally be taken first – the text used will be computer generated using a pre-recorded tape which will also contain voice announcements. Up to three candidates at a time will receive the same test. This will be followed by the sending test which will be taken individually.

The receiving test

In the receiving test the candidate is required to receive a minimum of 120 letters and seven figures in the form of a typical exchange between radio amateurs. The duration of the test will be approximately six minutes. Each character will be sent at a speed of 12wpm with a longer-than-normal gap between each character and word to reduce the overall reception speed to 5wpm. Each character incorrectly received will count as one error. A group of characters, which could include figures in which more than one character is received incorrectly, counts as two errors. If more than six errors are made, the test will result in failure. The candidate will not be permitted to write down the Morse symbols for later translation.

The sending test

In the sending test the candidate will be given a text to send by hand on a straight Morse key, consisting of not less than 75 letters and five figures. The text should be sent at not less than 5wpm; this should take approximately three minutes. The text will be in the form of a typical exchange between radio amateurs. There must be no uncorrected errors in the sending and not more than four corrections will be permitted.

The test could include any of the following Q-codes, commonly used abbreviations or procedural characters. Note that the procedural code 'CT' (commencing sign) will be sent right at the start of the receiving test but that this will not be part of the test for marking purposes.

Q-codes

QRA QRG QRK QRL QRM QRN QRP QRQ QRS QRT QRV QRX QRZ QSA QSB QSL QSO QSY QTH

Abbreviations

ABT	–	about
AGN	–	again
ANT	–	antenna
BK	–	signal used to interrupt a transmission in progress
CPI	–	copy
CPY	–	copy
CQ	–	general call to all stations
CUL	–	see you later
CW	–	continuous wave (Morse code)
DE	–	from, used to proceed the callsign of the calling station
DR	–	dear
EL	–	element
ES	–	and
FB	–	fine business
FER	–	for
GA	–	good afternoon
GD	–	good day
GE	–	good evening
GM	–	good morning

Notes

HPE	–	hope
HR	–	here
HVE	–	have
HW	–	how
K	–	invitation to transmit
MNI	–	many
MSG	–	message
NW	–	now
OC	–	old chap
OM	–	old man
OP	–	operator
PSE	–	please
PWR	–	power
R	–	received
RPRT	–	report
RST	–	readability, signal strength, tone report
RX	–	receiver
SIG	–	signal
SRI	–	sorry
TEMP	–	temperature
TKS	–	thanks
TNX	–	thanks
TU	–	thank you
TX	–	transmitter
UR	–	your
VERT	–	vertical
VY	–	very
WID	–	with
WX	–	weather
XYL	–	wife
YL	–	young lady
73	–	best wishes
88	–	love and kisses

Procedural characters and punctuation

$\overline{AR}$	· — · — ·	end of message
$\overline{CT}$	— · — · —	preliminary call
$\overline{BT}$	— · · · —	separation signal
$\overline{KN}$	— · — — ·	transmit, only the station called
$\overline{VA}$	· · · — · —	end of transmission
?	· · — — · ·	question
/	— · · — ·	oblique
Erase	· · · · · · · ·	to correct an error

Summary

The test details are summarised in the following table:

Test	Minimum number of characters	Approximate duration of test	Maximum number of errors	Speed of test
Sending	Letters 75 Figures 5	3 minutes	4 corrected	Not less than 5wpm
Receiving	Letters 120 Figures 7	6 minutes	6 uncorrected	Overall 5wpm with 12wpm character speed

Note that figures and procedural characters are counted as two letters for timing purposes.

Notes

EMC is the ability of electronic equipment to 'live together' without causing or receiving mutual interference.

If you hold a portable AM radio next to a computer or television set you will probably hear some strange noises such as buzzes and whistles. Under these circumstances the equipment could be said to be not electromagnetically compatible.

Sources of interference

- Radio transmitters in taxis, police cars, ambulances, as well as transmitters used by CB operators and radio amateurs.
- Other electrical equipment, particularly things which make sparks – thermostats and car ignition systems, electric motors in drills, razors, sewing machines etc.
- Interference from electrical equipment can cause spots on TV screens or buzzes, whines or clicks on TV or radio sound.
- Radio transmitters can also affect television receivers or video recorders, causing a loss of colour or a patterning on the picture. They can also cause a voice to break through on TV sound or broadcast radio. Although such problems could be the fault of the transmitter, they often occur because the radio or TV is receiving something which it isn't designed to receive.
- A radio transmitter can also affect equipment which shouldn't receive radio signals at all. For example, it could break through into a stereo system playing a record, cassette or CD, or into a telephone.

If any interference problems arise when you are licensed, the first thing to do will be to make sure that your amateur station is not producing any unwanted signals such as harmonics (see Worksheet 27). A transmitter should have a low-pass filter (as shown in Worksheet 26) to reduce harmonics.

A transmitter which is free of harmonics may still cause interference because the radio is not good enough at rejecting strong signals. This is more likely to happen with SSB than with CW, and least likely with FM. There are two approaches to solving this problem. The first would be to reduce the strength of the signal being sent to the affected equipment. This could be done in a number of ways:

- by keeping your transmitting aerial as high and as far away as possible from TV aerials, houses or wiring (mains or telephone).
- by keeping your aerial feeder cables away from house wiring. Screened cable should be of good quality, otherwise signals can leak out.
- by making sure that your transmitter is correctly matched to your aerial.
- by avoiding certain types of aerial, such as end-fed aerials.
- by using a good RF earth.
- by using a mains filter to prevent transmitted signals getting into the house wiring.
- by using only enough power necessary to make the contact.

The second way to solve the problem would be to cut down the amateur signals getting into the TV set or video recorder by plugging in a suitable high-pass filter at the aerial socket.

Social aspects

If a neighbour complained of any sort of interference you would need to help. Checking your log-book would show if you were 'on the air' at the

Notes

time of the interference and, if so, on which band. You would need to do tests to prove whether or not your transmissions actually caused interference. If they did, you would need to check all the points mentioned previously. If the TV required a filter it would be useful to have one handy which you could lend to the neighbour.

If you needed help you could contact your nearest RSGB EMC Co-ordinator (see the *RSGB Amateur Radio Call Book*). If your neighbour needed help, s/he should contact the dealer who sold the affected equipment. In the case of a telephone bought or rented from BT plc, s/he should contact BT.

A useful leaflet called *Advice on TV and Radio Reception* is available from the Radiocommunications Agency (phone 0171-211 0211). It also contains a form which the neighbour can use to ask the Radio Investigation Service (RIS) to visit if the dealer cannot solve the problem.

Part 6   Books and future plans

A long list of RSGB and other publications appears every month in the pages of *Radio Communication* and your instructor will be able to advise you of the books best suited to your needs. One or two are worth mentioning here.

- *Amateur Radio Operating Manual*, ed Ray Eckersley, G4FTJ
- The current edition of the *RSGB Amateur Radio Call Book*
- *D-i-Y Radio* magazine
- *Amateur Radio for Beginners (How to discover the hobby)*, Victor Brand, G3JNB
- *Practical Antennas for Novices*, John Heys, G3BDQ
- *The Radio Amateur's Guide to EMC*, Robin Page-Jones, G3JWI
- *Revision Questions for the Novice RAE*, Esde Tyler, G0AEC
- *Practical Transmitters for Novices*, John Case, GW4HWR
- *The Morse Code for Radio Amateurs*, George Benbow, G3HB

Useful background reading

- DTI booklet BR68a/N – *The Novice Licence*
- *A Closer Look at the Novice Licence*

These can be obtained from the RSGB – send a stamped and addressed envelope (A5) to the Amateur Radio Dept at RSGB HQ.

EMC

- *Breakthrough*
- *Use of Ferrite Rings*
- *Neighbour's Questions Answered*

All obtainable from RSGB HQ on receipt of a SAE.

Your instructor will encourage you to progress from the Novice Licence to a full licence and will tell you how to go about it. Useful books for this are the *Radio Amateurs' Examination Manual* and *How to Pass the RAE*, both by George Benbow, G3HB.

Circuit symbols

Description		Unit	Symbol
Resistor	– general	ohm	
	– variable		
	– potentiometer		
Capacitor	– general	farad	
	– polarised		
	– variable		
Inductor	– general	henry	
	– iron cored		
Transformer			
Crystal		hertz	
Semiconductor diode, general symbol			
Light emitting diode (LED)			
NPN transistor			
Cell		volts	
Battery		volts	
Loudspeaker			
Earphone			
Microphone			
Switch			
Fuse			
Lamp			
Antenna			
Earth			

(The above symbols are taken from British Standard BS 3939.)

The Novice Licence Schedule

Those licensed under an Amateur Radio (Novice) Licence (B) may not transmit on these bands between 1.950 and 28.500MHz.

Frequency bands (MHz)	Status of allocations in the United Kingdom to the Amateur Service	Maximum DC input (W)	Power RF output (W)	Permitted types of transmission
1.950–2.000	Available on the basis of non-interference to other services (inside or outside the United Kingdom)	5	3	Morse Telephony RTTY Data
3.560–3.585	Primary, shared with other services	5	3	Morse
10.13–10.14	Secondary	5	3	Morse
21.100–21.149	Primary	5	3	Morse
28.060–28.190	Primary	5	3	Morse RTTY Data
28.225–28.300	Primary	5	3	Morse RTTY Data
28.300–28.500	Primary	5	3	Morse Telephony
50.0–51.0	Primary. Available on the basis of non-interference to other services outside the United Kingdom. Antennas limited to 20m above ground level. No maritime mobile operation.	5	3	Morse Telephony Data
51.0–52.0	Secondary. Available on the basis of non-interference to other services outside the United Kingdom. Antennas limited to 20m above ground level. No maritime mobile operation.	5	3	Morse Telephony Data
432.00–435.00	Secondary	5	3	Morse Telephony Data
435.0–440.0	Secondary	5	3	Morse Telephony Data SSTV FSTV
1240–1325	Secondary	5	3	Morse Telephony RTTY Data Facsimile SSTV FSTV
10,000–10,500	Secondary	5	3	Morse Telephony RTTY Data Facsimile SSTV FSTV